SpringerBriefs in Molecular Science

History of Chemistry

Series Editor

Seth C. Rasmussen, Department of Chemistry and Biochemistry, North Dakota State University, Fargo, ND, USA

Springer Briefs in Molecular Science: History of Chemistry presents concise summaries of historical topics covering all aspects of chemistry, alchemy, and chemical technology. The aim of the series is to provide volumes that are of broad interest to the chemical community, while still retaining a high level of historical scholarship such that they are of interest to both chemists and science historians.

Featuring compact volumes of 50 to 125 pages, the series acts as a venue between articles published in the historical journals and full historical monographs or books.

Typical topics might include:

- An overview or review of an important historical topic of broad interest
- Biographies of prominent scientists, alchemists, or chemical practitioners
- New historical research of interest to the chemical community

Briefs allow authors to present their ideas and readers to absorb them with minimal time investment. Briefs are published as part of Springer's eBook collection, with millions of users worldwide. In addition, Briefs are available for individual print and electronic purchase. Briefs are characterized by fast, global electronic dissemination, standard publishing contracts, easy-to-use manuscript preparation and formatting guidelines, and expedited production schedules. Both solicited and unsolicited manuscripts are considered for publication in this series.

More information about this subseries at http://www.springer.com/series/10127

Kevin C. de Berg

The Iron(III) Thiocyanate Reaction

Research History and Role in Chemical Analysis

 Springer

Kevin C. de Berg
Avondale College of Higher Education
Cooranbong, NSW, Australia

ISSN 2191-5407   ISSN 2191-5415   (electronic)
SpringerBriefs in Molecular Science
ISSN 2212-991X
SpringerBriefs in History of Chemistry
ISBN 978-3-030-27315-6   ISBN 978-3-030-27316-3   (eBook)
https://doi.org/10.1007/978-3-030-27316-3

This Springer imprint is published by the registered company Springer Nature Switzerland AG
The registered company address is: Gewerbestrasse 11, 6330 Cham, Switzerland

Preface

Histories of chemistry and histories of science commonly focus on the historical development of ideas, themes, concepts and the personalities involved in these historical developments. Some histories will focus on a particular chemist giving insight into their family background, academic life and contributions made to chemistry or more broadly, scientific endeavour. Some examples that come to mind have been those written about Robert Boyle, Joseph Priestley, Antoine Lavoisier, Fritz Haber, Ernest Rutherford and Madame Curie just to name a few. It is less common to write a history about a chemical reaction but I have discovered that such a history will involve personalities, controversial ideas and applications to the broad landscape of science. By its very nature, a history of a chemical reaction will involve addressing some detailed chemistry, probably more than one might expect in a history focussed on a personality or a particular scientific idea. In the eighteenth century, chemists were disappointed that their discipline lacked the rigour that mathematics had brought to physics. It proved much easier to apply mathematics to visible objects like planets, projectiles and inclined planes than to the invisible constituents of matter undergoing chemical change. The development of calculus and its application to the thermodynamics and kinetics of chemical reactions proved an important turning point for chemistry. To demonstrate this often forgotten legacy, I do make use of mathematics in this book, but a level of mathematics not beyond that of a typical chemistry graduate. Some of the basic principles are revisited for the benefit of the reader where derivations are given. However, my challenge in a book series like this is to provide enough detail that will satisfy the historian and the chemist and I trust I will have been at least partly successful in achieving this.

My interest in history began in the mid-1980s when I completed a master's thesis on the history of the gas laws. What provoked my interest was the question: 'How did we come to express gas laws in the form used in modern chemistry'? This interest coincided with the establishment of the International History and Philosophy of Science and Science Teaching Group (IHPST) in 1989 and the opportunity to make subsequent contributions to its journal, *Science & Education-Contributions from History, Philosophy and Sociology of Science*, as an author and reviewer. In 2014, I contributed a chapter on the role of the history of chemistry in the teaching and learning of chemistry for the *International Handbook of Research in History, Philosophy and Science Teaching.*

The iron(III) thiocyanate reaction, where an intense blood-red colour is produced by mixing solutions of ferric ions and thiocyanate ions, has captured my interest since high school days. The reaction was used in my first-year university chemistry laboratory days as a test for the presence of ferric ions and when I was given the opportunity to teach senior high school chemistry for a time, the reaction was used as the centrepiece for equilibrium study in what was then known in the 1970s as the CHEM STUDY syllabus. More recently, I was given the opportunity to study the reaction in the laboratory of Professor Marcel Maeder at the University of Newcastle since there were still some unresolved issues with the reaction. Two papers, published in *Inorganica Chimica Acta*, have resulted from this collaboration to date.

One is often asked the question, particularly by fellow chemists: 'What use is a study of history to the chemistry community or more broadly to the scientific community, or even more broadly to the general community'? My own reflections on an answer to this question have to do with the contribution of history to our broad understanding of how knowledge of our world has been built through many circuitous paths involving errors, controversies and clashes of culture. Even while giving specific detailed study to a particular chemical reaction like the iron(III) thiocyanate reaction, these reflections remain with me, even though occasionally sitting in the background. Hopefully, such a study will make us sympathetic to the role science plays in providing information on broader controversial topics like climate change.

Cooranbong, Australia   Kevin C. de Berg

Acknowledgements

I wish to acknowledge four sources that have had some influence on my embarking on the study of a chemical reaction from an historical perspective.

1. Associate Professor Michael Matthews, founding president of the IHPST group, provided an international platform upon which newcomers to history and philosophy of science research could receive helpful reviews of initial attempts at writing in this area from experts in the field. These reviews helped to hone one's skills at writing history while still being involved in the practice of a science discipline, which in my case, was chemistry.
2. Avondale College of Higher Education provided funding towards my travel and presentation of a paper on the iron(III) thiocyanate reaction at the 11th International Conference for the History of Chemistry in Trondheim, Norway, in 2017.
3. Professor Marcel Maeder from the University of Newcastle willingly gave me access to his laboratory facilities for conducting further research on the iron (III) thiocyanate reaction and from his chemometrics background was able to suggest procedures for addressing some difficulties which proved valuable in the final analysis.
4. Dr. Sarah Clifford from the University of Newcastle who willingly gave me practical assistance in learning how to operate the stopped-flow apparatus and its computer interface for recording large amounts of absorbance data.

Contents

About the Author

Kevin C. de Berg completed his Ph.D. in physical chemistry in 1978 at the University of Queensland and his MAppSc in 1989 from Curtin University. His research thesis for the MAppSc degree examined the historical significance of the gas laws particularly for the education context. Interests in chemistry and history and philosophy of science led to an invitation to serve on the editorial committee of the journal: *Science & Education-Contributions from History, Philosophy, and Sociology of Science and Mathematics*; and to contribute a chapter titled: *The Place of the History of Chemistry in the Teaching and Learning of Chemistry*, for the International Handbook of Research in History, Philosophy and Science Teaching (2014). Kevin has spent 9 years as a High School Science and Mathematics Teacher and 36 years lecturing and researching in the area of physical and inorganic chemistry and history and philosophy of science. He is currently Conjoint Associate Professor at Avondale College of Higher Education.

Chapter 1
Introduction

1.1 Eighteenth Century Background

Readers familiar with the iron(III) thiocyanate reaction from their chemistry laboratory experience at college or university will most likely have used the reaction as a
test for the presence of iron(III) or thiocyanate ions in chemical analysis. Iron(III)
in the presence of thiocyanate ions forms a blood-red coloured solution which was
the identification trigger. The childish excitement that I experienced when I saw this
happen is captured by Knight [1] when discussing chemistry as a service science:

> Bubblings, test-tubes becoming too hot to hold, sharp colour changes, sudden turbidities and
> precipitations are the stuff of chemistry: a science where the secondary qualities of things
> (ideas linked to sense data) are of great importance if one is to get a sense of what is going
> on in reactions.

As far as we know, the iron(III) thiocyanate reaction first came to the attention of
chemists in the early 19th century, but the context in which it appeared had its origins
in the 17th and 18th centuries. Duncan [2] summarizes what chemists of the 18th
century thought about their subject as follows.

> To sum up, chemists of the mid-eighteenth century-and later-presented an earlier stage of
> chemistry as having been mystical, obscure, and deluded, or as consisting merely of experi-
> ments without an organizing theory. They further presented the chemistry of their own era as
> having emerged at some time in the later seventeenth or early eighteenth century from that
> obscurity, and having become a rational, organized science, with a theoretical basis, on a par
> with such mathematically based branches of natural philosophy as astronomy or Newtonian
> mechanics. At the same time, they insisted that theory must be based on experiment and not
> imagined *a priori*.

Certainly the precision and exactness of Newtonian mechanics was something that
chemists of the 18th century were hoping to achieve but whether they were successful in achieving this at that stage is arguable. While chemistry had its origins
in alchemy, by the beginning of the 18th century chemists had generally become
known as artisans of an applied science. On the other hand, physicists had become
known as natural philosophers of an exact science which was based on precision,

© The Author(s), under exclusive license to Springer Nature Switzerland AG 2019
K. C. de Berg, *The Iron(III) Thiocyanate Reaction*, SpringerBriefs in
History of Chemistry, https://doi.org/10.1007/978-3-030-27316-3_1

quantification, mathematics and theory based on experiment and observation rather than speculation. It was these attributes of an exact science that chemists wanted to emulate in the 18th century and beyond but this was not to occur without considerable difficulty.

Henry [3] dates the Scientific Revolution from 1543, the time of the publication of Copernicus' *On the Revolutions of the Heavenly Spheres*, to around 1704 by which time Isaac Newton (1643–1727) had published his *Principia Mathematica* (1687) and *Opticks* (1704). He assesses Newton's contribution at the end of the Scientific Revolution and the beginning of the European-wide Enlightenment in these terms [3]:

> ….the end of the Scientific Revolution was effectively marked by the triumphant work of Isaac Newton. In just two land-mark books, Newton seemed to bring the major lessons of the Scientific Revolution together and to demonstrate the new power of the physical sciences. His *Principia Mathematica* of 1687 was seen as an unassailable demonstration of how mathematics could be used to help an understanding of the workings of the physical world, while the *Opticks* of 1704 was seen as the perfect exemplar of the experimental method.

Berlin [4] was to confirm "Newton's influence as the strongest single factor" in understanding the 18th century Enlightenment. In the preface to his *Principia* [5], Newton was to alert chemists to the difficulties they would face in trying to understand the nature of substances and their chemical changes with the same rigour with which he was able to treat the earth and its planets.

> We derive from celestial phenomena the gravitational forces by which bodies tend toward the sun and toward the individual planets. Then the motions of the planets, the comets, the moon, and the sea are deduced from these forces by propositions that are also mathematical. If only we could derive the other phenomena of nature from mechanical principles by the same kind of reasoning! For many things lead me to have a suspicion that all phenomena may depend on certain forces by which the particles of bodies, by causes not yet known, either are impelled toward one another and cohere in regular figures, or are repelled from one another and recede. Since these forces are unknown, philosophers have hitherto made trial of Nature in vain. But I hope that the principles set down here will shed some light on either this mode of philosophizing or some truer one.

John Freind (1675–1728) was a chemist who attempted to reduce chemistry to the 'roots of true philosophy' which was the Newtonian philosophy of attraction and repulsion between particles of matter which varied in their shape, texture and density [6]. James Crawford (1682–1731) criticised this approach as a hypothesis not based on experimental evidence [7]. While the inverse square law worked for visible objects like the moon and the planets, on what basis could we say it also applied to particles we cannot observe? Pierre Joseph Macquer (1718–1784), the great French chemist responsible for compiling a chemical dictionary, also considers the application of what he calls the 'geometrical method' to minute invisible particles on which chemical phenomena depends as unwarranted [8]. It might work for astronomy, but not necessarily for chemistry.

The intimate relationship between theory and experiment was accepted uncritically by most chemists of the 18th century. Herman Boerhaave (1668–1738), a teacher of medicine and chemistry in Leyden, always used the principle of collecting as much data as possible before devising an explanation using a theoretical model [9]. Macquer [10] makes a similar point:

> Theory can be useful only as far as it is born from experiments already carried out, or as it shows us those which are to be carried out. ....If experiment which is in no way directed by theory is always a blind groping, theory without experiment is never more than a deceptive and undependable glimpse. Thus it is certain that the most important discoveries which have been made in chemistry are due only to the combination of these two great resources.

But when is a theory simply speculation and when is it not? Was Robert Boyle (1627–1691) within the bounds of acceptable practice to postulate that acids consisted of particles with spikes and alkalis consisted of particles with smooth holes to accommodate these spikes even though these particles were not visible? Was the fact that acids have a sour taste and alkalis have a smooth feel sufficient to warrant the postulate? Duncan [11] comes to the conclusion that, "In any age, what is counted as being an obvious inference from the results of experiment and what is counted as being wild speculation is very much a matter of convention."

As to whether chemistry had in fact reached that stage of maturity described as an exact science by the end of the 18th century, it is helpful to confer with Antoine Lavoisier (1743–1794) [12], the great French chemist some repute to be the father of chemistry, "Perhaps one day the precision of the data might be brought to such perfection that the mathematician in his study would be able to calculate any phenomenon of chemical combinations in the same way, so to speak, as he calculates the movement of the heavenly bodies." It would be another 100 years before chemistry could approach the sophistication of physics and astronomy. While physicists of the 18th century were happy to make the

> optimistic assumption that the unseen workings of nature followed tidy, mechanical, material patterns, capable ultimately of being expressed quantitatively in the Newtonian manner....most chemists...considered that the complexities of chemistry made the link too difficult to discover, and that perhaps the link was not a simple or direct one [13].

While chemists did not achieve the ultimate goal in the 18th century, they did prepare and characterise many new substances upon which 19th century chemistry would be able to build.

Aristotle's four elements—earth, air, fire and water—and the later three scholastic principles—mercury, sulphur, and salt—filtered into the texts of 18th century chemistry but it was never clear exactly what the differences between the elements and the principles were. Boerhaave [14] still believed that metals consisted of the philosophical principles—sophic mercury and sophic sulphur. The 18th century use of the elements and principles is well summarised by Duncan [15]:

> ...the traditional notions of the four Aristotelian elements and the three philosophical principles were modified in practice so that the elements and principles were understood by chemists to be distinct substances, which remained more or less unchanged during chemical reactions. Even after 1700 there was still some vagueness about the conception of them. ....However, the traditional elements and principles bore less and less relationship to the substances actually identified and named in laboratories. ....Eventually, however, having ceased to be useful, the traditional notions of elements and principles ceased to be mentioned and the concept of elements derived from chemical practice replaced them.

The prime example of this move to a practical definition of 'element' was Lavoisier [16]:

> I shall therefore only add upon this subject, that if, by the term elements, we mean to express those simple and indivisible atoms of which matter is composed, it is extremely probable we know nothing at all about them; but, if we apply the term elements, or principles of bodies, to express our idea of the last point which analysis is capable of reaching, we must admit, as elements, all the substances into which we are capable, by any means, to reduce bodies by decomposition. Not that we are entitled to affirm, that these substances we consider as simple may not be compounded of two, or even of a greater number of principles; but, since these principles cannot be separated, or rather since we have not hitherto discovered the means of separating them, they act with regard to us as simple substances, and we ought never to suppose them compounded until experiment and observation has proved them to be so.

Lavoisier's cautiousness about the constitution of matter is exemplary for the time. Even though there was a ready acceptance of the particulate nature of matter by both physicists and chemists, evidence for the shape and texture of these particles was not forthcoming and it was not until John Dalton's (1766–1844) atomic theory of the early 19th century that chemists would have any chance of quantifying the mass of these particles.

The attempt to use particle shape and texture to explain the nature of chemical reactions proved a failure in the 18th century. The fact that substance A reacted with substance B in preference to substance C was explained in terms of A being more strongly *attracted* to B than C, or A having a stronger *affinity* for B than C. However, the reason for the preference for B over C was not clear. Some chemists thought A and B must have been similar in some way, such as composition, but Joseph Black (1728–1799) [17] observed that "affinity implies, or suggests, some similarity which, in most cases, is not agreeable to fact, seeing that we generally observe the greatest dissimilarity in those bodies which are eminently prone to unite." While Black preferred the term *attraction* or *chemical attraction* to distinguish it from *gravitational attraction*, others preferred the term *affinity* or in the case of Étienne François Geoffroy (1672–1731) [18] and his affinity table, *rapports*, but eventually all chemists understood the use of all the terms to mean 'tendency to react'.

The construction of Tables of Affinity during the 18th century helped to organise the large variety of chemical reactions and substances in a form which was at least partly helpful to the chemist. For example, metal displacement reactions known in the 18th century could be organised into a column with the following metals going in order from top to bottom: iron, tin, lead, copper, silver, mercury, gold. This means that iron can displace tin from a solution of its salts, tin can displace lead from a solution of its salts, lead can displace copper from a solution of its salts, and so on. In terms of the concept of *affinity*, one could express the relationships in the column as: iron has a stronger affinity for the salts of tin than tin has for the salts of iron, tin has a stronger affinity for the salts of lead than lead has for the salts of tin, and so on. But such tables had their critics like Antoine Grimoald Monnet (1734–1817) [19] who said that, "substances act on each other according to the state in which they happen to be, rather than according to their respective affinities….and the system of affinities is a beautiful chimera, better fitted to amuse our scholastic chemists than to advance that science." Monnet's point is a pertinent one although chemists did find some benefit from the Affinity Tables. The difficulty in achieving consistency in orders of affinity eventually led to their demise in the 19th century. However, in 1855 the concept of *affinity* provided the context for introducing the iron(III) thiocyanate reaction and Bergman's concept of *complete reaction* and Berthollet's concept of *incomplete reaction* [20]. Torbern Olaf Bergman (1735–1784) was a Swedish chemist and mineralogist noted for his 1775 *Dissertation on Elective Attractions* which contained his extensive chemical affinity tables. Claude Louis Berthollet (1748–1822) was a French chemist known for his contribution to the theory of chemical equilibrium and his contribution to the new nomenclature for naming chemical compounds.

1.2 Publication Summary of the Iron(III) Thiocyanate Reaction Over 191 Years

A representative sample of publications related to the iron(III) thiocyanate reaction from 1826 to 2017 is shown in Table 1.1. This period of 191 years covers some of the most interesting and significant episodes in the history of chemistry. Upon the death of Lavoisier in 1794 it was the Swedish chemist Jöns Jacob Berzelius (1779–1848) who continued the challenge of developing a symbolic representation for elements and compounds on the basis of the initial letter or letters of the Latin name. Thus was given the symbol, Fe, for iron from the Latin, ferrum. According to Sir Harold Hartley [21], "As a systematist Berzelius was the natural successor to Lavoisier and completed much that Lavoisier had begun." While Lavoisier eschewed speculation and theory in favour of observation and experiment (although Lavoisier was not able to discuss the results of many of his experiments without recourse to the caloric theory), Berzelius was ready to grasp the essential ideas behind Dalton's atomic theory and the law of definite proportions and pursue rigorous laboratory methods to determine a table of atomic weights based on the oxygen atom which was given a value of 100. His analytical ability also led to the determination of the formula for thousands of compounds and the discovery of new elements like cerium (1803), selenium (1818) and thorium (1828). It was during his analysis of iron compounds that he observed what we now call the iron(III) thiocyanate reaction.

A scan over the entries in Table 1.1, representing 191 years of iron(III) thiocyanate chemistry, reveals some important ideas which have a broad significance for chemistry beyond the iron(III) thiocyanate reaction itself. These are as follows: composition and the development of nomenclature, formulae and units; qualitative and quantitative chemical analysis; issues related to chemical affinity and chemical equilibrium; the identification of the molecular or ionic species responsible for the blood-red colour; the role of mathematics and data analysis in determining equilibrium and rate data; and improvements in instrumentation particularly the spectrophotometer. Historians have attempted to divide the history of chemistry into epochs,

Table 1.1 A summary of the iron(III) thiocyanate system as it is manifested in the chemistry literature from 1826 to 2017

Year	Author/s/reference	Features
1826	Berzelius [22]	Identification of an intense red colour on mixing thiocyanic acid with an iron compound
1855	Gladstone [20]	Reaction: $Fe_2Rd_3 + 3M,S_2Cy = Fe_2,3S_2Cy + 3MRd$ Reaction used to decide between Bergman and Berthollet's view of a reaction
1885	Thomson [23]	Uses the iron(III) thiocyanate reaction to determine small quantities of iron often present in other compounds. No formulae given and no discussion of the reaction
1907	Stokes and Cain [24]	Uses the iron(III) thiocyanate reaction to determine small quantities of iron. Attributes the red colour to ferric sulphocyanate and its double compounds although no formulae given. A mixture of amyl alcohol and ether was used to extract most of the iron as sulphocyanate. Attributes the fading of an aqueous solution of the ferric sulphocyanate to the reduction of iron to the ferrous state by isodisulphocyanic acid $(HN:CS)_2$. Addition of $K_2S_2O_8 + Hg(SCN)_2.2HSCN$ designed to stop oxidation of HSCN and reduction of ferric ions
1913	Roscoe and Schorlemmer [25]	Formula given for ferric thiocyanate: $Fe(SCN)_3.3H_2O$. Bleaching of the colour due to reduction of iron to the ferrous state. Oxidation products of thiocyanate thought to be CO_2, H_2SO_4, and NH_3. Double salts known to be $M_4Fe(CNS)_6$ and $M_3Fe(CNS)_6$
1913	Philip and Bramley [26]	Reaction written as: $FeCl_3 + 3KCNS \rightleftharpoons Fe(CNS)_3 + 3KCl$. Displacement of equilibrium can be detected by changes in colour intensity. Paper focuses on the oxidation products of thiocyanate and suggests CO_2, SO_4^{2-}, NH_3, and HCN. Reduction rate $\uparrow$ as $[CNS^-] \uparrow$
1924	Bailey [27]	Reaction written as: $FeCl_3 + 3KSCN \rightleftharpoons Fe(SCN)_3 + 3KCl$ $Fe(SCN)_3 \rightleftharpoons Fe^{...} + 3(SCN)^{///}$ When $Fe(SCN)_3$ is removed by ether, more forms when the ions combined as a result of disturbing the equilibrium because the reaction is subject to the law of mass action. No evidence for dithiocyanate in solution

(continued)

Table 1.1   (continued)

Year	Author/s/reference	Features
1927	Bailey [28]	Red colour not due to the peroxidised ferrous salt, $FeHC_3N_3S_3O_3$, suggested by Tarugi, but most likely ferric thiocyanate (presumably $Fe(SCN)_3$)
1931	Schlesinger and Van Valkenburgh [29]	Red colour in aqueous solution due to $[Fe(CNS)_6]^{3-}$. Red colour of ether extracts due to $Fe[Fe(CNS)_6]$. Based on molecular weight determination in ether of $2 \times Fe(CNS)_3$; red colour migrating to anode and ferric ions to the cathode
1937	Fowles [30]	Used the reaction in the form: $FeCl_3 + 3NH_4CNS \rightleftarrows Fe(CNS)_3 + 3NH_4Cl$, to assess the impact of changing concentrations on the rates of the forward and reverse reactions
1941	Woods and Mellon [31]	Reviews the thiocyanate method for determining iron. Beers Law (absorbance proportional to concentration) obeyed up to 10 ppm iron. Nature of the reaction remains controversial-some attribute the colour to $Fe(SCN)_6^{---}$, some to $Fe(SCN)^{++}$, others to $Fe(SCN)^+$. As $[SCN^-] \uparrow$, absorbance shifts from 460 to 480 mμ. Using 60% acetone decreases fading and increases stability of the colour. Thiocyanate method for iron is inferior to other methods
1941	Bent and French [32]	Found evidence for $FeSCN^{2+}$ but not $Fe(SCN)_3$ or $Fe(SCN)_6^{3-}$; $\mu = 0.665$; $[FeCl_3] = 0.003582$ M; $[NaSCN]$ from 0.000109 to 0.6322 M. Higher complexes possible but difficult to calculate. Uses: $\log Fe_m(CNS)_n = m \log Fe^{3+} + n \log CNS^- - \log K$ to plot $\log [CNS^-]$ against $\log$ (optical density) to find $n = 1$ in the dilute case. $K = 0.033 \pm 0.002$ and $\varepsilon = 7880 \pm 550$ at 500 mμ
1941	Edmonds and Birnbaum [33]	Confirmed the colour due to $FeSCN^{2+}$. Considered $Fe^{+++} + nCNS_{b-nx}^- \rightleftarrows Fe(CNS)_{nx}^{3-n}$ $a - x$ and showed that $(a - x)(b - nx)^n = (b - x)(a - nx)^n$ which can only be true when $n = 1$. Found K by solving $(a - x)b^n/x = K$ and $(a - y)d^n/y = K$ simultaneously
1942	Gould and Vosburgh [34]	Applied Job's method (equal concentrations of iron and thiocyanate used to form different ratio mixtures) to the iron(III) thiocyanate system. Confirmed $FeSCN^{2+}$ being a single compound of low stability

(continued)

Table 1.1 (continued)

Year	Author/s/reference	Features
1947	Frank and Oswalt [35]	Confirmed the colour due to $FeSCN^{2+}$ measuring absorbances from 4000 to 5000 Å. Used a series expansion of a quadratic to determine K and ε. For $\mu = 0.5$, $K_1 = 138.0 \pm 1.9$ corrected for hydrolysis. $\varepsilon = 4700 \pm 2\%$ and λ_{max} is 4475 Å. Determine $\log K_1^0$ to be 2.95 from the D-H law. Observed λ_{max} to shift to lower energy as $[SCN^-] \uparrow$. Claimed no evidence of fading
1949	Polchlopek and Smith [36]	Applied Job's method for Fe:SCN ratios of 1:1 to 1:6. The complex $Fe(SCN)_2^+$ indicated in the 0.5–1 M conc. range. Measurements made immediately due to instability of complexes. Observed λ_{max} to shift to lower energy as $[SCN^-] \uparrow$ again observed. No attempt to find K given the complexity of the system
1950	Harvey and Manning [37]	Observed pronounced fading after 30 min at 460 mµ. Using excess conc. of 40×10^{-4} M and non-excess conc. of 1×10^{-4} M to 6×10^{-4} M they used the slope ratio method to get n = m = 1 for A_mB_n and the Edmonds-Birnbaum equation to find $K = 58.8$ at $\mu = 0.6$
1951	Macdonald et al. [38]	Studied the distribution of ferric thiocyanate between ether and water. The possibility of getting complexes from $FeSCN^{2+}$ up to $Fe(SCN)_6^{3-}$ was assumed. Solns unstable and measured as quickly as possible-more stable in ether. Used the Frank & Oswalt eqn. to find ε and K for $FeSCN^{2+}$ of 4983 ± 40 at 4500 Å and 127.9 ± 6 at $\mu = 1.0$ M respectively. Used the method of least squares to estimate K_2 to K_6. Measurements made from 3900 to 5500 Å
1953	Betts and Dainton [39]	Gives attention to the spontaneous bleaching of ferric thiocyanate acidified solutions. SCN^- converts Fe^{3+} to Fe^{2+} and forms $(SCN)_2$ which hydrolyzes to give $SCN + CN^- + SO_4^{2-}$. Simultaneous eqns using $K_1 = D/a(bE-D)$ used to find K_1 and E. For $\mu = 1.28$ at 18.8 °C, $K_1 = 120.4 \pm 0.5$ and $E = 5000 \pm 50$. Rabinowitch-Stockmeyer eqn. used to calculate $\log K_1^0 = 2.94$. Kinetics used to calculate $K_2 = 20 \pm 5$

(continued)

Table 1.1 (continued)

Year	Author/s/reference	Features
1953	Lewin and Wagner [40]	Useful summary of data to 1953: red colour due to $FeSCN^{2+}$ and $Fe(SCN)_2^+$ well established from spectrometric and conductometric studies; $Fe(SCN)_3$ or $Fe_2(SCN)_6$ in organic solvents. From migration studies higher complexes could be present
1955	Lister and Rivington [41]	Serious fading for Fe^{3+} 0.1 M and SCN^- 10^{-4} M due to oxid. of SCN^- by oxygen carried by Fe^{3+}-prevented by adding benzyl alcohol. When SCN^- was high conc. fading alleviated with H_2O_2 as different cause. Frank & Oswalt eqn. used to find ε and K. ε_1 at 4600 Å is 4680. D-H law used to find $K_0 = 1070$. Successive approximations used to find $K_2 = 15.5 \pm 1.0$ at 25 °C and $\mu = 1.2$ with ε_2 of 9130 at 4800 Å
1956	Laurence [42]	A potentiometric study of the reaction. Used the method of successive approximations to find $K_1 = 139 \pm 0.5$ and $K_2 = 20.9 \pm 1$ at 25 °C and $\mu = 0.5$. Used the D-H law in the form used by Rabinowitch and Stockmeyer to determine $K_1^0 = 1070$ using data from other sources. No mention made of colour fading
1958	Perrin [43]	Focus on $Fe(SCN)_2^+$. Used spectrophotometric and potentiometric methods. Fading a problem and had to take readings within 30 min of mixing. Spectro. gave $K_1 = 145$, $K_2 = 14$ at 18 °C and $\mu = 0.56$. Potent. gave $K_1 = 133$, $K_2 = 10$ at 20 °C and $\mu = 0.65$. $[SCN^-]$ found by successive approximations; had to use published value of ε_1 to get K_1. ε_2 was found to be 9800 at 4850 Å. K_1^0 and K_2^0 determined to be 1090 and 40 respectively. Only 1:1 and 1:2 complexes up to 0.15 M SCN^-. $Fe(SCN)_2^+$ major absorbing species when $SCN^- > 0.04$ M
1958	Below et al. [44]	Study of the kinetics of formation of $[Fe(H_2O)_5SCN]^{2+}$ using a rapid mixing device, the forerunner of stopped-flow. Suggests a possible 3-step mechanism: $$Fe(H_2O)_6^{3+} \rightleftarrows Fe(H_2O)_5(OH)^{2+} + H^+$$ $$Fe(H_2O)_5(OH)^{2+} + SCN^- \rightleftarrows Fe(H_2O)_4(OH)(SCN)^+ + H_2O$$ $$Fe(H_2O)_4(OH)(SCN)^+ + H^+ \rightleftarrows Fe(H_2O)_5(SCN)^{2+}$$

(continued)

Table 1.1 (continued)

Year	Author/s/reference	Features
1961	Yalman [45]	In the process of discussing the stability of $FeSCNF^+$, use spectrophotometric and potentiometric methods along with successive approximations and the Frank-Oswalt eqn. to find K_1 (Potent.) $= 141$ and K_2 (Potent.) $=13.2$ at 26.7 °C and $\mu = 0.5$; K_1 (Spectro.) $= 142.5$ at 26.7 °C and $\mu = 0.52$ and $\varepsilon_1 = 4621$ (at λ?)
1963	Ramette [46]	Interested in $FeSCN^{2+}$. Adds successive portions of (0.1 M Fe^{3+} + 0.5 M H^+) to (2 × 10^{-4} M SCN^- + 0.5 M H^+). Attributes colour fading to light. Uses the Frank-Oswalt eqn. to find $K = 169$ at $\mu = 0.5$ and $\lambda_{max} = 445$–450 nm. Temp. not given
1964	Carmody [47]	Demonstrates Job's method of continuous variations for confirming the formula, $FeSCN^{2+}$, the 1:1 formula
1972	Goodall et al. [48]	Relaxation kinetics for formation of $Fe(H_2O)_5(SCN)^{2+}$ using temperature-jump and flash photolysis techniques. Finds support for the 3-step mechanism proposed in 1958 by Below et al.
1977	Sammour et al. [49]	Stability of iron(III) thiocyanate complexes. Found the order of complexation of Fe^{3+} with anions other than SCN^- was $SO_4^{2-} > Cl^- > NO_3^- > ClO_4^-$
1979	Driscoll [50]	Approaches the Fe^{3+}/SCN^- equilibrium from an enquiry perspective. Considers the reaction as: $FeCl_3 + 3NH_4CNS = Fe(CNS)_3 + 3NH_4Cl$; mixes equal volumes of 0.001 M $FeCl_3$ and 0.001 M NH_4CNS and divides the mixture into 5 parts and compares colours to a standard when solids are added to each part
1980	Gray and Workman [51]	Follows the kinetics using stopped-flow for 1000 ms for production of $Fe(H_2O)_5(SCN)^{2+}$. Same reaction steps used as for Goodall et al. [48] and Below et al. [44]. Ratio of rate constants gives $K = 120$ M^{-1} at 25 °C and $\mu = 1.0$ M. No mention made of colour fading
1982	Sultan and Bishop [52]	Only found 1:1 complex even when SCN^- was in large excess. Noted decomposition of complex observable within a minute due to reduction of Fe^{3+} by SCN^-. Got unusual values for: $K = 1.284$ M^{-1} and $\varepsilon = 1458$. Temp. and ionic strength not mentioned

(continued)

Table 1.1 (continued)

Year	Author/s/reference	Features
1983	Funahashi et al. [53]	Mechanism of iron(III) complexation with SCN^- up to high concentrations. Used high pressure stopped-flow at 25 °C. Found the formation of the 1:1 complex rate determining and formation of higher complexes fast. The reaction was first order with respect to Fe^{3+} and SCN^-
1985	Bjerrum [54]	Distribution of tris(thiocyanato)iron(III) between octan-2-ol and aqueous thiocyanate solutions. For $\mu = 1.0$ M, $\log K_1 = 2.30 \pm 0.05$, $\log K_2 = 1.69 \pm 0.05$, $\log K_3 = 0.4 \pm 0.1$. Although higher $\log K$'s were estimated, they are not reliable. $Fe(SCN)_3$ is the only species extracted into the alcohol. Eleven simultaneous eqns. solved by a computer program. Complexes decompose slowly-more stable in the alcohol. SCN^- normally coordinates to Fe^{3+} through N
1986	Hill and MacCarthy [55]	Shows the connection between Job's method of continuous variations and titrimetry. 30 mL of 0.001 M $Fe(NO_3)_3$ + 0.01 M HNO_3 was titrated with 0.001 M KSCN + 0.015 M HCl. Attributes the instability of the complex to exposure to light
1987	Grompone [56]	Used the iron(III) thiocyanate reaction to determine the amount of iron in a bar of soap. Solutions used were 0.105 M NH_4SCN and iron(III) concs up to 5.37×10^{-5} M for calibration
1992	Ozutsumi et al. [57]	Connected a titration vessel to a flow cell with path length 0.5 cm. Used 50 λ's between 350 nm and 650 nm. Minimised the square of the difference between calc. and expt. values. Up to 0.1 M SCN^- and $\mu = 1.0$ M found $K_1 = 128.8$ and $\varepsilon_1 = 4500$; $K_2 = 16.98$ and $\varepsilon_2 = 8700$; $K_3 = 3.02$ and $\varepsilon_3 = 9000$. Attributes the instability to light exposure so expt. done in dark. Uses a non-linear least squares algorithm proposed by Marquardt
1995	Echols and Tyson [58]	Used a flow injection procedure giving a K higher than the published data. Acknowledges this inaccuracy but says procedure instructive. No mention of instability of complex

(continued)

Table 1.1 (continued)

Year	Author/s/reference	Features
1997	Clark [59]	Stopped-flow kinetics of formation of Iron(III) Thiocyanate. Similar reactions considered to previous kinetic investigations. K_f (kinetics) = 138 ± 29 at 25 °C and $\mu = 1.0$ M. No mention of colour fading
1998	Cobb and Love [60]	One mL portions of 0.1 M Fe^{3+} + 0.5 M $HClO_4$ added to 100 mL of 0.0002 M SCN^- + 0.5 M $HClO_4$ then mixed, placed in cuvette to measure absorbance then returned to the 100 mL beaker and continued like this. Frank-Oswalt eqn. used. Davies eqn. used for K^0. At $\mu = 0.5$ M $K = 146$ and log $K^0 = 2.94$. No mention of colour fading
1999	Stolzberg [61]	50 mL 0.1 M $Fe(NO_3)_3$ + 50 mL 0.001 M NaSCN and volume made up to 1 L. Then divided into 10/75 mL lots and different amounts of KNO_3 added to see the effect on K. Frank-Oswalt eqn. used. The idea was to challenge students into whether one can argue that K is really a constant—looking at the impact of ionic strength without mentioning the concept
1999	Lahti et al. [62]	Determination of thiocyanate in human saliva using the iron(III) thiocyanate system using 0.2 M $Fe(NO_3)_3$ and 2×10^{-4} M KSCN for standard curve
2011	Najib and Hayder [63]	Claim to have observed complexes from 1:1 to 1:6 when adding SCN^- to 8×10^{-4} M Fe^{3+} up to about 6.5 M SCN^-. Claim that the instability of the complexes due to inner redox reaction between SCN^- and CNS^- and add $KMnO_4$ to prevent this
2011	Nyasulu and Barlag [64]	Sequential addition of one reagent to the other in a cuvette with stirring. 0.1 mL of 0.001 M KSCN added to 4 mL 0.2 M $Fe(NO_3)_3$ in a cuvette and 0.5 mL 0.002 M KSCN added to 4 mL 0.002 M $Fe(NO_3)_3$. Calibration method or Frank-Oswalt method used
2014	Ghirardi et al. [65]	Used Gladstone's [20] experiment showing incomplete reaction as part of a teaching/learning sequence for equilibrium study

(continued)

Table 1.1 (continued)

Year	Author/s/reference	Features
2016	de Berg et al. [66]	Stopped-flow used to determine initial spectra at time zero to handle the instability of the complexes. React-Lab equilibria and React-Lab kinetics were used to work with 77,000 spectra to determine $K_1 = (98 \pm 1)\text{M}^{-1}$ and $K_2 = (7 \pm 1)\text{M}^{-1}$ at 25 °C and $\mu = 0.5$ M. Values are lower than published ones due to using the spectra at time zero
2017	de Berg et al. [67]	Initial spectra used to determine the impact of ionic strength on K. The D-H law in the form of the Specific Ion Interaction theory used to find $\log K_1^0 = (2.85 \pm 0.08)$ and $\log K_2^0 = (1.51 \pm 0.13)$. The changes in the empirical specific ion interaction coefficients were $\Delta\varepsilon_1 = (-0.29 \pm 0.16)$ and $\Delta\varepsilon_2 = (-0.18 \pm 0.25)$

Symbols used in the table have the following meaning: μ ionic strength; K equilibrium constant; K^0 equilibrium constant at zero ionic strength; ε (or E) molar absorptivity; D-H Debye–Hückel law; λ_{max} wavelength at maximum absorbance; ms milliseconds; mμ millimicron; nm nanometre; Å angstroms; subscripts 1 and 2 refer to complexes ML and ML$_2$ respectively where M = Fe^{3+} and L = SCN$^-$

stages or eras so that chemistry can be viewed in a broad sense as part of the history of ideas. The German chemist and historian Hermann Kopp [68] divided chemistry into the Ancient Stage (Pre-4th century AD), the Age of Alchemy (350–1525 AD), the Age of Medicinal Chemistry (1525–1650), the Phlogiston Period (1650–1775), and the Age of Quantitative Investigation (Post-1775). Many of the ideas in Table 1.1 certainly fit the Quantitative period but more recent attempts have focused on considering stages within the post-1775 period given that more than 170 years have elapsed since Kopp's analysis.

Jensen [69] divided the post-1775 period into three periods of chemical revolution: the First Chemical Revolution (1770–1790) which coincided with the development of a new nomenclature for chemical species and a new understanding of combustion reactions; the Second Chemical Revolution (1855–1875) which coincided with the determination of atomic and molecular weights and the emergence of the concepts of valence and molecular structure; and the Third Chemical Revolution (1904–1924) which highlighted the important role of the electron in chemistry including electron structure in atoms and molecules.

Historians recognize that classification periods in the history of chemistry are not meant to have rigid boundaries. In regard to the suggested three periods of chemical revolution, Jensen [69] reminds us that, "Classification is based on abstracting certain characteristics and overlooking others. Consequently it almost always ignores a certain inherent fuzziness in its categorical boundaries that results from the overlap of one class with another." To illustrate this we have seen that Berzelius and his chemistry, the first entry in Table 1.1, lies between the period of Jensen's first and second chemical revolutions, completing the work begun by Lavoisier on the new chemical nomenclature and engaging with Dalton's atomic theory to develop a table of atomic weights. From the abstracted boundaries of chemical revolution one gets the impression that issues of composition and nomenclature would have been resolved by the beginning of the third chemical revolution. A scan through Table 1.1 confirms there was still an issue of composition, nomenclature and formula in 1855 for the components of the iron(III) thiocyanate reaction which seemed to be resolved by the beginning of the 20th century only to be revised again by the middle of the 20th century owing to the emergence of the field of coordination chemistry. Thus, according to Table 1.1, iron(III) thiocyanate, the product of adding together ferric ions and thiocyanate ions, appeared as: ferric sulphocyanide, $Fe_2,3S_2Cy$, in 1855; ferric thiocyanate, $Fe(CNS)_3$, in 1913; and pentaaquathiocyanatoiron(III), $[Fe(H_2O)_5SCN]^{2+}$, in 1958.

Improvements in instrumentation, particularly absorption spectrophotometry as far as the iron(III) thiocyanate reaction is concerned, have proved vital in reaching an understanding of a reaction that has been with us since 1826 according to Table 1.1. Initially, at the beginning of the 20th century, the comparison of colours between a sample colour and a standard colour was with the naked eye looking through two vertical tubes and adjusting the level of liquid until the colours matched. By the middle of the 20th century with the development of electronics, this system was replaced with a machine that could measure the intensity of light emerging from a coloured solution and thus recording a transmittance value and an absorbance value. But by the end of the 20th century with the assistance of the modern computer, the absorption spectrophotometer was a smaller streamlined instrument that could perform tasks only dreamed of 100 years earlier. Instrumentation appears in Jensen's third chemical revolution in the form of spectroscopy but this period of revolution ended around 1924. Improved computing facilities along with sophisticated data analysis packages have exponentially improved the capacity of instrumentation since then. The developments since 1924 recorded in Table 1.1 for the iron(III) thiocyanate reaction bear witness to this. Instrumentation is the focus of a fourth chemical revolution (1945–1966) proposed by Chamizo [70] and while this highlights instrumentation to a greater extent than Jensen's classification, it doesn't account for the growth in the way computers have been made to interface with an instrument which in turn is interfacing with a chemical entity such as an electron or a nucleus.

To be fair to Jensen and Chamizo, their classifications focus on chemical entities rather than on applications that use the properties of these entities for analysis purposes, and so modern analysis techniques are not a focus of their classifications. Jensen focuses on the molar composition of matter or the macroscopic features of matter, molecular structure or microscopic features of matter, and finally electron structure or the sub-microscopic. Chamizo focuses on the atom, the molecule, the electron and the nucleus, and spin. In addition, they consider these entities were arguably part of what Kuhn [71] would call revolutionary chemistry as opposed to normal chemistry. Because of their orientation neither mentions the role of mathematics in producing some of the laws and theories of chemistry and this becomes paramount when studying the nature of a chemical reaction like the iron(III) thiocyanate reaction. Since the 'nature of a chemical reaction' would have to be at the 'heart' of chemistry, there may be grounds for considering an alternative classification of chemistry to those that have been proposed, even though they contain some useful ideas. Could one not argue that discovering the 'heart' of chemistry was revolutionary rather than normalcy?

The nature of the iron(III) thiocyanate reaction is such that it easily lends itself to a discussion of some philosophical ideas like laws, theories and models and also what philosophers call epistemic thinking, and these ideas will be taken up in this book. It is very difficult to talk about history of chemistry without some interface with philosophy.

It is interesting to observe from the data in Table 1.1 that chemical analysis for iron occurred as early as 1885, well before chemists had identified the species responsible for the blood-red colour, the chemical equation for the reaction, and its equilibrium constant. During the early days of the Royal Society of Chemistry, membership was dominated by analysts who were not particularly interested in the ultimate nature of matter or reactions [72]. It would seem appropriate then to dedicate Chap. 2 to the role of the reaction in chemical analysis before undertaking the more theoretical task of understanding the nature of the reaction.

References

1. Knight D (1992) Ideas in chemistry. The Athlone Press, London, p 178
2. Duncan A (1996) Laws and order in eighteenth-century chemistry. Clarendon Press, Oxford, p 5
3. Henry J (2015) Science in the Athens of the North: the development of science in enlightenment Edinburgh. In: Anderson RGW (ed) Cradle of chemistry. John Donald, Edinburgh, p 8
4. Berlin I (ed) (1956) The age of enlightenment: the eighteenth-century philosophers. New American Library, New York, p 15
5. Newton I (1999) The Principia: mathematical principles of natural philosophy (trans: Cohen IB, Whitman A). University of California Press, Berkeley, CA, pp 382–383
6. Freind J (1712) Chymical lectures, in which almost all the operations of Chymistry are reduced to their true principles, and the laws of nature. Phillip Gwillim for John Bowyer, London
7. Crawford J (1713) Tractatus Chymici (a Doctore Crawford Dictati). MS 2451, Wellcome Library, London
8. Macquer PJ (1766) Dictionnaire de chymie, vol I. Paris
9. Boerhaave H (1701) Oratio de Commendando Studio Hippocratio. Leiden
10. Macquer PJ (1766) Dictionnaire de chymie, vol I. Paris, pp xxii–xxiii
11. Duncan A (1996) Laws and order in eighteenth-century chemistry. Clarendon Press, Oxford, p 75
12. Lavoisier A (1785) Oeuvres, vol II. In: Dumas JB, Grimaux E, Fouqué FA (eds) (6 vols) (1862–1893), Impériale, Paris, p 550
13. Duncan A (1996) Laws and order in eighteenth-century chemistry. Clarendon Press, Oxford, p 85

14. Boerhaave H (1741) New method of chemistry (trans of Elementa chemiae vol II by Shaw P). London
15. Duncan A (1996) Laws and order in eighteenth-century chemistry. Clarendon Press, Oxford, pp 8–9
16. Lavoisier A (1789) Traité Élementaire de Chimie (trans: Kerr R 1790). Edinburgh, pp xii-i–xxxvii
17. Black J (1803) Lectures on the elements of chemistry. In: Robinson J (ed), vol I. Edinburgh, pp 266–267
18. Geoffroy EF (1719) Table des différens rapports observés en chimie entre différentes substances 1718. Mémoires de l'Académie Royale des Sciences: 202–212
19. Monnet AG (1775) Traité de la dissolution des métaux. Amsterdam and Paris, p 55
20. Gladstone JH (1855) On circumstances modifying the action of chemical affinity. Phil Trans R Soc Lond 145:179–223
21. Hartley HH (1971) Studies in the history of chemistry. Clarendon Press, Oxford, p 136
22. Berzelius JJ (1826) Lehrbuch der Chemie. Arnold, Dresden
23. Thomson A (1885) Colorimetric method for determining small quantities of iron. J Chem Soc Trans 47:493–497
24. Stokes HN, Cain JR (1907) On the colorimetric determination of iron with special reference to chemical reagents. J Am Chem Soc xxix(4):409–447
25. Roscoe HE, Schorlemmer C (1913) A treatise on chemistry vol II-the metals, 5th edn. Macmillan & Co. Ltd., London
26. Philip JC, Bramley A (1913) The reaction between ferric salts and thiocyanates. J Chem Soc 103:795–807
27. Bailey KC (1924) The reaction between ferric chloride and potassium thiocyanate. Proc Roy Irish Acad B 37:6–15
28. Bailey KC (1927) Ferric thiocyanate. J Chem Soc cclxxiii:2065–2069
29. Schlesinger HI, Van Valkenburgh HB (1931) The structure of ferric thiocyanate and the thiocyanate test for iron. J Am Chem Soc 53:1212–1216
30. Fowles G (1937) Lecture experiments in chemistry. G Bell & Sons Ltd., London
31. Woods JT, Mellon MG (1941) Thiocyanate method for iron. A spectrophotometric study. Ind Eng Chem 13(8):551–554
32. Bent HE, French CL (1941) The structure of ferric thiocyanate and its dissociation in aqueous solution. J Am Chem Soc 63:568–572
33. Edmonds SM, Birnbaum N (1941) Ferric thiocyanate. J Am Chem Soc 63:1471–1472
34. Gould RK, Vosburgh WC (1942) Complex ions III. A study of some complex ions in solution by means of the spectrophotometer. J Am Chem Soc 64:1630–1634
35. Frank HS, Oswalt RL (1947) The stability and light absorption of the complex ion $FeSCN^{++}$. J Am Chem Soc 69:1321–1325
36. Pokhlopek SE, Smith JH (1949) Composition of ferric thiocyanate at high concentrations. J Am Chem Soc 71:3280–3283
37. Harvey AE, Manning DL (1950) Spectrophotometric methods of establishing empirical formulas of coloured complexes in solution. J Am Chem Soc 72:4488–4493
38. Macdonald JY, Mitchell KM, Mitchell ATS (1951) Ferric thiocyanate. Part II. The distribution of ferric thiocyanate between ether and water. J Chem Soc: 1574–1580
39. Betts RH, Dainton FS (1953) Electron transfer and other processes involved in the spontaneous bleaching of acidified aqueous solutions of ferric thiocyanate. J Am Chem Soc 75:5721–5727
40. Lewin SZ, Wagner RS (1953) The nature of iron(III) thiocyanate in solution. J Chem Educ 30(9):445–449
41. Lister MW, Rivington DE (1955) Some measurements on the iron(III)-thiocyanate system in aqueous solution. Can J Chem 33(10):1572–1590
42. Laurence GS (1956) A potentiometric study of the ferric thiocyanate complexes. Trans Far Soc 52:236–242
43. Perrin DD (1958) The ion $Fe(CNS)_2^+$. Its association constant and absorption spectrum. J Am Chem Soc 80(15):3852–3856

44. Below JF, Connick RE, Coppel CP (1958) Kinetics of the formation of the ferric thiocyanate complex. J Am Chem Soc 80:2961–2967
45. Yalman RG (1961) Stability of the mixed complex $FeSCNF^+$. J Am Chem Soc 83:4142–4146
46. Ramette RW (1963) Formation of monothiocyanatoiron(III). J Chem Educ 40(2):71–72
47. Carmody WR (1964) Demonstrating Job's method with colorimeter or spectrophotometer. J Chem Educ 41(11):615–616
48. Goodall DM, Harrison PW, Hardy MJ, Kirk CJ (1972) Relaxation kinetics of ferric thiocyanate. J Chem Educ 49(10):675–678
49. Sammour HM, Sheglila AJ, Ny FA (1977) Stability of iron(III)-thiocyanate complexes and the dependence of absorbance on the nature of the anion. Analyst 102:180–186
50. Driscoll DR (1979) Invitation to enquiry: the Fe^{3+}/CNS^- equilibrium. J Chem Educ 56(9):603
51. Gray ET, Workman HJ (1980) An easily constructed and inexpensive stopped-flow system for observing rapid reactions. J Chem Educ 57(10):752–755
52. Sultan SM, Bishop E (1982) A study of the formation and stability of the iron(III)-thiocyanate complex in acidic media. Analyst 107:1060–1064
53. Funahashi S, Ishihara K, Tanaka M (1983) Mechanism of iron(III) complex formation. Activation volumes for the complexation of the iron(III) ion with thiocyanate ion and acetohydroxamic acid. Inorg Chem 22:2070–2073
54. Bjerrum J (1985) The iron(III)-thiocyanate system. The stepwise equilibria studied by measurements of the distribution of tris(thiocyanato)iron(III) between octan-2-ol and aqueous thiocyanate solutions. Acta Chem Scand A 39:327–340
55. Hill ZD, MacCarthy P (1986) Novel approach to Job's method. J Chem Educ 63(2):162–167
56. Grompone MA (1987) Determination of iron in a bar of soap. J Chem Educ 64(12):1057–1058
57. Ozutsumi K, Kurihara M, Miyazawa T, Kawashima T (1992) Complexation of iron(III) with thiocyanate ions in aqueous solution. Anal Sci 8:521–526
58. Echols RT, Tyson JF (1995) Determination of formation quotients by a flow injection procedure. Analyst 120:1175–1179
59. Clark CR (1997) A stopped-flow kinetics experiment for advanced undergraduate laboratories: formation of iron(III) thiocyanate. J Chem Educ 74(10):1214–1217
60. Cobb CL, Love GA (1998) Iron(III) thiocyanate revisited. A physical chemistry equilibrium lab incorporating ionic strength effects. J Chem Educ 75(1):90–92
61. Stolzberg RJ (1999) Discovering a change in equilibrium constant with change in ionic strength. J Chem Educ 76(5):640–641
62. Lahti M, Vilpo J, Hovinen J (1999) Spectrophotometric determination of thiocyanate in human saliva. J Chem Educ 76(9):1281–1282
63. Najib FM, Hayder OI (2011) Study of stoichiometry of ferric thiocyanate complex for analytical purposes including F^- determination. Iraqi Nat J Chem 42:135–155
64. Nyasulu F, Barlag R (2011) Colorimetric determination of the iron(III)-thiocyanate reaction equilibrium constant with calibration and equilibrium solutions prepared in a cuvette by sequential addition of one reagent to the other. J Chem Educ 88(3):313–314
65. Ghirardi M, Marchetti F, Pettinari C, Regis A, Roletto E (2014) A teaching learning sequence for learning the concept of chemical equilibrium in secondary school education. J Chem Educ 91(1):59–65
66. de Berg K, Maeder M, Clifford S (2016) A new approach to the equilibrium study of iron(III) thiocyanates which accounts for the kinetic instability of the complexes particularly observable under high thiocyanate concentrations. Inorg Chim Acta 445:155–159
67. de Berg K, Maeder M, Clifford S (2017) The thermodynamic formation constants for iron(III) thiocyanate complexes at zero ionic strength. Inorg Chim Acta 446:249–253
68. Kopp H (1843) Geschichte der Chemie, vol 1. Vieweg, Braunschweig
69. Jensen WB (1998) Logic, history, and the chemistry textbook III. One chemical revolution or three? J Chem Educ 75(8):961–969
70. Chamizo JA (2019) About continuity and rupture in the history of chemistry: the fourth chemical revolution (1945–1966). Found Chem 21:11–29

71. Kuhn T (1970) The structure of scientific revolutions, 2nd edn. University of Chicago Press, Chicago
72. Knight D (1992) Ideas in chemistry. The Athlone Press, London, pp 158–159

Chapter 2
The Reaction and Chemical Analysis

It is interesting to observe in Table 1.1 that the iron(III) thiocyanate reaction was used very early in chemical analysis, often without any knowledge of the reaction profile or formulae for reactants and products. This is testament to the experimental skill of our pioneer chemists. According to Knight [1], when reviewing the state of chemistry in the early 19th century, "For every experiment in chemistry done to test some theory, there must be at least 999 done to see what something is made of." And as far as the techniques were concerned [1]:

> Eminent analysts like Berzelius and Liebig developed qualitative and quantitative techniques. Reactions were sometimes done in the 'dry way', notably when substances were heated with borax on a charcoal block, using a blowpipe (the method of breathing while blowing had to be learned); but usually in the 'wet' or 'humid' way, in solution. There was no adequate theory: that had to wait for the Law of Mass Action and other developments in physical chemistry. The important thing was therefore to follow methods which gave consistent and reproducible results, and to guard against any known sources of error.

We will first consider qualitative analysis and then address quantitative analysis.

2.1 Qualitative Analysis

Berzelius [2] takes note in his observations of the reactions of iron salts that, "…ist auch diese Säure eins der empfindlichsten Reagentien für Eisen…" (this acid is also one of the most sensitive reagents for iron) because of the "rothe färbt" (red colour) produced. The acid he was referring to was thiocyanic acid (HSCN) or what he called, "schwefelcyanwasserstoffsäure". This test is still used as a qualitative test for ferric ions and thiocyanate ions.

In my first-year laboratory chemistry course at university, we were given an aqueous solution mixture containing at least one of the ions, Fe^{3+}, Co^{2+}, and Ni^{2+}, and had to test for their presence or absence. Ammonia solution was first added to the mixture: if Fe^{3+} was present a precipitate of $Fe(OH)_3$ would form, whereas cobalt and nickel ions would form the hexaammine soluble compounds. To positively confirm

© The Author(s), under exclusive license to Springer Nature Switzerland AG 2019   19
K. C. de Berg, *The Iron(III) Thiocyanate Reaction*, SpringerBriefs in
History of Chemistry, https://doi.org/10.1007/978-3-030-27316-3_2

the dirty orange-brown precipitate as being a ferric compound, we were asked to dissolve the precipitate in dilute HCl and add ammonium thiocyanate whereupon a blood-red colour would confirm the presence of the ferric ion. Students in our current first-year chemistry class are asked to perform identification tests for anions and one of the tests for the nitrite ion, NO_2^-, is to add thiourea to an acidified solution of the nitrite ion and then add an aqueous solution of ferric chloride, $FeCl_3$, after which an intense blood-red colour appears. The reaction sequence is:

$$CS(NH_2)_2 + HNO_2 \rightarrow N_2 + HSCN + 2H_2O$$
$$Fe^{3+} + SCN^- \rightarrow \text{blood-red colour}$$

2.2 Quantitative Analysis

While the iron(III) thiocyanate reaction has been used for a long time as a qualitative test for the presence of either ferric ions or thiocyanate ions because of the blood-red colour that forms on mixing ferric ions with thiocyanate ions, attempts to use the reaction as a quantitative test for iron were made as early as 1885 [3] (see Table 1.1). The purity of many compounds at this time was compromised often by the presence of small quantities of iron and other metal ions and Thomson's study was designed to show how the quantity of iron could be determined by the formation of iron(III) thiocyanate. It is interesting to note that Thomson does not write a chemical equation for the reaction and does not discuss the formula of the product of the reaction and his analysis occurred before the invention of the spectrophotometer in 1940.

The technique involved using two glass cylinders, A and B, into which solutions of the unknown, dilute acid, and thiocyanate were added to A and just the dilute acid and thiocyanate to B. Distilled water was added to both cylinders to the 100 mL mark and then a standard solution of iron(III) was added to B until the colours in A and B visually matched. The standard iron solution was 1.785×10^{-3} M and the potassium thiocyanate solution was 0.4116 M. Without a knowledge of the chemical reaction and the associated equilibrium constants, one could not be sure if the thiocyanate concentration was high enough to react with all the iron present, so tests had to be done with solutions of known iron concentration to see what correlation existed between the known iron concentration and that calculated from the colour matching experiment. The ratio of the calculated to the known concentrations varied from 1:0.92 to 1:1 and on to 1:1.07. After considering possible interferences from the presence of other metal ions, Thomson [4] concluded, "From its almost universal applicability, simplicity, accuracy, and the great celerity with which it can be applied, this process for estimating small quantities of iron should be found extremely useful."

In 1907, Stokes and Cain [5] (Table 1.1) used a similar if not somewhat more elaborate colorimetric technique than used by Thomson in 1885. The technique again involved comparing the colours in two vertical tubes, one containing standard iron and the other the unknown iron. However, Stokes and Cain drew attention to the fact, not always reported, that the blood-red colour fades over time and they expended some effort in attempting to eliminate this fading. Thomson had noted in 1885 that the presence of mercuric salts and oxalic acid bleach the blood-red colour, but did not report the gradual fading that occurs even in the absence of these bleaching agents. Gladstone [6] had earlier reported the bleaching effect of the presence of mercuric salts and oxalic acid in 1855. Stokes and Cain attributed the fading to the reduction of the ferric ion to the ferrous ion by isodisulphocyanic acid, $(HN:CS)_2$, which, they claim, always forms when sulphocyanates (thiocyanates) are acidified. They found the use of the double compound, $Hg(SCN)_2.2HSCN$, prevented the oxidation of HSCN and a few milligrams of potassium persulphate prevented the reduction of ferric ion. In addition, a mixture of isoamyl alcohol and ether improved the extraction of the blood-red compound so they could report [7], "We thus obtain a solution of iron in a great excess of free sulphocyanic acid, practically free from all other substances, and so secure the most favourable conditions possible for the complete conversion of the iron into undissociated ferric sulphocyanate." As far as the analysis was concerned, 5 mL of amylic mixture (5:2 isoamyl alcohol-ether by volume), 5 mL sulphocyanic reagent ($Hg(SCN)_2.2HSCN$), 5 mL water, and a few milligrams of potassium persulphate were added to a test cylinder and a standard cylinder. A known volume of unknown iron solution was added to the test cylinder and sufficient standard iron solution added to the standard cylinder to give the same colour intensity as in the test cylinder, determined using the naked eye.

By 1941 photoelectric spectrophotometers had been built to determine light intensity transmitted or absorbed in the visible and ultraviolet range through solutions. Woods and Mellon [8] (Table 1.1) summarized the thiocyanate method for determining iron up to this point. They concluded that the nature of the reaction leading to the blood-red colour remained controversial with some attributing the colour to $Fe(SCN)_6^{--}$, some to $Fe(SCN)^{++}$, and others to $Fe(SCN)^+$. They found 60% acetone in water decreased the fading of the colour and increased its stability. The Beer-Lambert law (absorbance proportional to concentration) was obeyed up to 10 ppm iron and the maximum absorbance peak shifted from 460 to 480 mμ as the thiocyanate concentration increased. A study of the interference effects of other ions and the difficulty associated with the fading of the colour led Woods and Mellon to conclude that the thiocyanate method for iron determination was inferior to other methods such as that depending on complexation of Fe^{2+} with o-phenanthroline.

However, the thiocyanate method persisted in analytical circles to the extent that Baily [9] could say in 1957, "The thiocyanate method for determining iron is one of the most convenient and is generally accepted as official in analytical practice." Baily found that using a mixture of 1 volume of acetone with 2 volumes of methyl ethyl ketone inhibited the fading of the colour for at least 90 min. As for all previous determinations of iron, the thiocyanate concentration had to exceed the iron concentration significantly if all the iron was to be complexed to give the blood-red colour. In Baily's procedure the thiocyanate concentration was about 1150 times that of the iron concentration.

Standard inorganic analysis texts written around 1960 [10, 11] refer to the thiocyanate method for determining iron. Both of these texts suggest two methods: the method of comparing colours in two cylinders like that used as early as 1885 by Thomson [3] and 1907 by Stokes and Cain [6] and the method of measuring absorbance at 480 mμ in a colorimeter. Vogel [12] suggested that the species responsible for the red colour will depend on the thiocyanate concentration: "At low thiocyanate concentrations the predominant coloured species is $Fe(SCN)^{2+}$, at 0.1 M thiocyanate concentration it is largely $Fe(SCN)_2^+$, and at very high thiocyanate concentrations it is $Fe(SCN)_6^{3-}$." Belcher and Nutten [13] attributed the colour "to the formation of $Fe(CNS)_3$, $Fe\{Fe(CNS)_6\}$ and $FeCNS^{2+}$." Both texts recommended measurement of absorbance immediately on mixing the reagents with Vogel [14] providing the reason: "since the colour fades on standing." By the 1960's, the species $Fe(SCN)^{2+}$ had definitely been confirmed in the research literature and $Fe(SCN)_2^+$ was also most likely formed at high thiocyanate concentrations. Some doubt still existed about the species $Fe(SCN)_6^{3-}$ with $Fe(CNS)_3$ being the most likely form in an organic medium.

In 2005, Achar and Bellappa [15] reported on a sensitive microspectrophotometric determination of iron by the thiocyanate method. They found 60% acetone stabilized the colour for more than three days as well as increasing the sensitivity of the method as reported also by Woods and Mellon [8] above. The thiocyanate concentration was in excess of the iron concentration by a factor of 25,000 to achieve maximum absorbance. A calibration curve of absorbance against concentration of iron was constructed and the unknown iron concentration determined from the graph using its absorbance. Neither Baily, nor Achar and Bellappa, give discussion to the reaction leading to the colour or to what species might be responsible for the colour.

The thiocyanate method has been readily adopted by chemistry educators. One of the purposes of quantitative analysis in chemistry classes has been to introduce students to important applications of chemistry in industry and pathology. The *Journal of Chemical Education* often features such applications and attention is drawn here to two such applications: one determining the amount of iron in a bar of soap by Grompone [16] (Table 1.1), and the other determining the amount of thiocyanate in human saliva by Lahti et al. [17] (Table 1.1). In the case of iron in soap, reference is made to the action of sodium hydroxide on the fat of the soap, but the iron determination is made without any reference at all to the iron(III) thiocyanate reaction or to the identity of the coloured species. This, of course, has been typical of analytical procedures. In the case of small amounts of iron, the assumption is made that all the iron will be converted to the blood-red coloured species in the presence of a relatively higher concentration of thiocyanate. Grompone [16] reported that the standard solutions contained 0.105 M ammonium thiocyanate and iron(III) concentrations varying from 8.95×10^{-6} to 5.37×10^{-5} M. Thus the thiocyanate concentration was of the order of 11,500 times that of the iron concentrations which is comparable to that used by analytical chemists. The unknown iron concentration was determined from a calibration curve of iron concentrations against absorbance.

In the case of the determination of thiocyanate in human saliva, reference was made to the iron(III) thiocyanate reaction to show that in the presence of excess iron one can conclude that the only species contributing to the blood-red colour will be $FeSCN^{2+}$ since thiocyanate complexes $Fe(SCN)_2^+$ and $Fe(SCN)_3$ can be excluded under these conditions. Lahti et al. [17] drew upon information that had been published by 1999 concerning the identity of iron(III) thiocyanate species. The standard solutions contained 0.1 M iron(III) nitrate and thiocyanate concentrations varied from 2×10^{-5} to 1×10^{-4} M. This meant that the iron concentration was in excess of the thiocyanate concentration by around a factor of 5000. Again, a calibration or standard curve was constructed of absorbance against thiocyanate as $FeSCN^{2+}$ concentration and the unknown value of thiocyanate in saliva determined from the standard curve. Apparently thiocyanate ion is a detoxification product of the reaction between cyanide and thiosulphates in the liver. Exposure to tobacco smoke can increase saliva thiocyanate concentration from a non-smoker range of 0.5–2 mM to about 6 mM in smokers. Metabolism of vitamin B12 also increases saliva thiocyanate concentration.

Up until the mid-1970's clinical pathology laboratories used the iron(III) thiocyanate reaction to determine total iron in lysated blood. Typically, a sample of heparinised blood was treated with concentrated sulphuric acid, potassium persulphate and sodium tungstate solution to ensure that the extracted iron existed as Fe^{3+}. On addition of 20% potassium thiocyanate the blood-red colour formed and the solution was made up to a set volume with deionized water ready for analysis. An iron solution containing 0.1 mg Fe^{3+} per mL was used to prepare solutions with a range of iron concentrations and the absorbance of each solution measured at around 540 nm so as to construct a calibration curve. This curve was used to convert the absorbance of the blood sample to an iron concentration. The test now used relies on iron existing as iron(II) and its binding to a multidentate ligand like ferrozine to produce a very stable complex [18]. This method overcomes the instability problem with iron(III) thiocyanate.

Did the fact that excess iron had to be used for thiocyanate determination, and excess thiocyanate had to be used for iron determination, register anything about the nature of the reaction in the chemist's mind? This might be hard to tell given that equilibrium data is rarely, if at all, referred to by chemists or educators when discussing chemical analysis.

Over the 191 year period covered by the information in Table 1.1, nomenclature, formulae, and units related to the iron(III) thiocyanate reaction underwent changes from 1826 to 2017. We will need to address these changes in Chap. 3 before we investigate the nature of the reaction used in chemical analysis in some detail.

References

1. Knight D (1992) Ideas in chemistry. The Athlone Press, London, p 159
2. Berzelius JJ (1826) Lehrbuch der Chemie. Arnold, Dresden, p 771
3. Thomson A (1885) Colorimetric method for determining small quantities of iron. J Chem Soc Trans 47:493–497

4. Thomson A (1885) Colorimetric method for determining small quantities of iron. J Chem Soc Trans 47:497
5. Stokes HN, Cain JR (1907) On the colorimetric determination of iron with special reference to chemical reagents. J Am Chem Soc xxix(4):409–447
6. Gladstone JH (1855) On circumstances modifying the action of chemical affinity. Phil Trans R Soc Lond 145:179–223
7. Stokes HN, Cain JR (1907) On the colorimetric determination of iron with special reference to chemical reagents. J Am Chem Soc xxix(4):411
8. Woods JT, Mellon MG (1941) Thiocyanate method for iron. A spectrophotometric study. Indust Eng Chem 13(8):551–554
9. Baily P (1957) Stabilization of ferric thiocyanate color in aqueous solution. Anal Chem 29(10):1534–1536
10. Belcher R, Nutten RJ (1960) Quantitative inorganic analysis, 2nd edn. Butterworths, London
11. Vogel AI (1962) A textbook of quantitative inorganic analysis, 3rd edn. Longmans, London
12. Vogel AI (1962) A textbook of quantitative inorganic analysis, 3rd edn. Longmans, London, p 785
13. Belcher R, Nutten RJ (1960) Quantitative inorganic analysis, 2nd edn. Butterworths, London, p 325
14. Vogel AI (1962) A Textbook of quantitative inorganic analysis, 3rd edn. Longmans, London, p 787
15. Achar BN, Bellappa S (2005) A modified sensitive micro spectrophotometric determination of iron(III) by thiocyanate method. Indian J Pharm Sc 67(1):119–122
16. Grompone MA (1987) Determination of iron in a bar of soap. J Chem Educ 64(12):1057–1058
17. Lahti M, Vilpo J, Hovinen J (1999) Spectrophotometric determination of thiocyanate in human saliva. J Chem Educ 76(9):1281–1282
18. Stookey LL (1970) Ferrozine-A new spectrophotometric reagent for iron. Anal Chem 42(7):779–781

Chapter 3
The Reaction and Its Nomenclature, Formulae, and Units

3.1 Nomenclature and Formulae

New compositional nomenclature had been proposed by four famous French chemists, Guyton de Morveau, Lavoisier, Fourcroy, and Berthollet in the late eighteenth century. Leicester [1] summarises their method of chemical nomenclature as follows: "The names of simple substances should express their characters when possible, and the names of compounds should indicate their composition in terms of their simple constituents." Thus 'oxygen' meant 'acid former' and 'ferric oxide' was a compound of iron and oxygen. This was in contrast to 'dephlogisticated air' for oxygen and 'the calx of iron' for ferric oxide. The committee set up by the French Academie des Sciences to review the new nomenclature expressed some reservation given the fact that the nomenclature appeared to be based on the 'oxygen theory' or 'antiphlogistic theory'. In reviewing the committee's report, Crosland [2] observes:

> Whereas many experiments were put forward in support of the latter (antiphlogistic theory), was it not also true, they (the committee) said, that the phlogiston theory was supported by a series of convincing experiments? The old phlogiston theory was no doubt incomplete, but were there not also some difficulties in the new theory?

The fact that even the French Academie was not comfortable in dismissing outrightly the phlogiston theory resonates with Hasok Chang's [3] claim that the transition from the phlogiston theory to the oxygen theory was by no means a smooth process and chemistry's progress during the nineteenth century would have been richer and more productive if the two models had been allowed to coexist for a little longer.

In 1855, formulae for salts were written quite differently to the way they are written today. Iron(III) nitrate was called ferric nitrate but its identifying formula was $Fe_2O_3, 3NO_5$ according to Gladstone [4]. This was because it was prepared by mixing a basic oxide, "ferric oxide (or the sesquioxide of iron)", Fe_2O_3, with an acidic oxide, nitric acid, NO_5. Reorganising the symbols one can see that Fe_2O_3, $3NO_5$ is not equivalent to the modern formula, $Fe(NO_3)_3$. One can see here that the issue relates to the formula used for nitric acid. In the mid-nineteenth century acids were still commonly regarded as oxygen carrying species following Lavoisier's

© The Author(s), under exclusive license to Springer Nature Switzerland AG 2019
K. C. de Berg, *The Iron(III) Thiocyanate Reaction*, SpringerBriefs in
History of Chemistry, https://doi.org/10.1007/978-3-030-27316-3_3

idea and any hydrogen associated with acidic properties was thought to come from water. So typical formulae for acids did not show hydrogen. At this time atomic weights had not been standardized so the formula for an acid depended on what set of atomic weights was used in the formula calculation. In 1808, Dalton [5] proposed the formula NO_2 for nitric acid but as the century progressed other oxides of nitrogen were proposed such as NO_5 as we have seen [4] and N_2O_5 [6]. According to modern chemistry, inorganic acids can be formed by dissolving a non-metal oxide in water. An example of this which leads to our current formula for nitric acid is, $N_2O_5 + H_2O \rightarrow 2HNO_3$. If Gladstone had used Fe_2O_3, $3N_2O_5$ to represent ferric nitrate, then this formula would have been equivalent to $2Fe(NO_3)_3$ showing our currently understood formula for ferric nitrate.

The foregoing paragraph highlights the difficulties chemists had in determining formulae for acids and consequently salts based on their composition according to the new French standards for writing formulae for compounds. A compositional formula showing the relative proportions of the elements was strongly criticised by the English chemist James Keir and the Irish physician Stephen Dickson. The basis of their objection concerned the then difficulties analysts had in determining accurate compositions [7]. Partly for this reason, some chemists preferred the old scheme of naming a compound based on its properties. Thus, *oil of vitriol*, was preferred over SO_3 (H_2SO_4–H_2O), the formula for sulfuric acid used in the nineteenth century. Nomenclature based on composition was eventually to win the day particularly with the eventual acceptance of Dalton's atomic theory. The numbers showing the proportion of elements could be interpreted as the number of atoms in the formula and chemical equations could be written so as to conserve the number of atoms.

What we know today as potassium thiocyanate was called the sulphocyanide of potassium in 1855 with formula K,S_2Cy [4]. According to Ihde [8], 'Cy' represented a combination of carbon and nitrogen as 'CN' although Gladstone doesn't use this translation. The red colour formed on mixing ferric nitrate with the sulphocyanide of potassium was called ferric sulphocyanide of formula, $Fe_2,3S_2Cy$, by Gladstone [9] who wrote the equation for the reaction as follows:

$$Fe_2Rd_3 + 3M, S_2Cy = Fe_2, 3S_2Cy + 3M\ Rd \tag{3.1}$$

where Rd represented a radical, a group of non-metal atoms, and M represented a metal atom. The 'equal' sign indicated conservation of atoms. If ferric nitrate was represented as Fe_2O_3, $3NO_5$, then Rd would represent NO_6 so the equation would be written, with CN replacing Cy, as:

$$Fe_2(NO_6)_3 + 3K, S_2CN = Fe_2, 3S_2CN + 3KNO_6 \tag{3.2}$$

If ferric nitrate was represented as Fe_2O_3, $3N_2O_5$, and Cy represented cyanogen, C_2N_2, then the equation could correspond to our current nomenclature (apart from the equals sign) as:

$$2Fe(NO_3)_3 + 6KSCN = 2Fe(SCN)_3 + 6KNO_3\ or$$

$$Fe(NO_3)_3 + 3KSCN = Fe(SCN)_3 + 3KNO_3 \qquad (3.3)$$

According to Table 1.1, the red compound was called ferric sulphocyanate in 1907 by Stokes and Cain [10] and the more familiar ferric thiocyanate in 1913 by Philip and Bramley [11], with the equation written showing forward and reverse arrows to indicate the reversibility of the reaction:

$$FeCl_3 + 3KCNS \rightleftarrows Fe(CNS)_3 + 3KCl \qquad (3.4)$$

With the advent of the ionic theory in the late 19th and early 20th centuries and the use of partial forward and reverse arrows to indicate equilibrium in 1902 [12], the equation was written in ionic form in 1924 [13] as a net ionic equilibrium equation where the dots are equivalent to positive charges and the dashes to negative charges for 3SCN:

$$Fe^{\cdots} + 3(SCN)^{///} \rightleftharpoons Fe(SCN)_3 \qquad (3.5)$$

By 1937, the ions were written as Fe^{3+} and SCN^- [14]. The partial forward and reverse arrows indicates incomplete transformation in both the forward direction and the reverse direction at equilibrium. With coordination chemistry emerging as a sub-discipline in the early 20th century, a new nomenclature was needed to name its compounds and chemists had to distinguish between *primary valency* (charge number) and *secondary valency* (coordination number). So, while Fe^{3+} can be regarded as having a primary valency of 'three', in aqueous solution it had a secondary valency of 'six' since six water molecules could coordinate to the Fe^{3+} ion to give, $Fe(H_2O)_6^{3+}$, with the name hexaaquairon(III). Consequently, the 1:1 complex was written as $Fe(H_2O)_5(SCN)^{2+}$ with the name pentaaquathiocyanatoiron(III), the 1:2 complex as $Fe(H_2O)_4(SCN)_2^+$ with the name tetraaquadithiocyanatoiron(III), and the 1:3 complex as $Fe(H_2O)_3(SCN)_3$ with the name triaquatrithiocyanatoiron(III) and so on. These are the forms in which the complex is written in the reaction mechanisms shown from 1958 in Table 1.1. When the red species was identified as a complex, the question which chemists needed to resolve was which of the complex species was responsible for the red colour. This question is taken up later in the book.

3.2 Units

In Eq. (3.1), Rd referred to a salt radical and M referred to a metal and the equation was meant to imply that one equivalent of Fe_2Rd_3 would react with three equivalents of MS_2Cy. In 1855, there was still much scepticism about Dalton's atomic theory, with many chemists preferring the experimental basis of equivalent weights over the more theoretical basis of atomic weights, the equivalent weight of an element being the weight of an element that would react with 1 g of hydrogen, 8 g of oxygen, or 35.5 g of chlorine. The problem with equivalent weights of elements was that they could be variable depending upon which reaction one was considering. In the case of compounds the equivalent weight was determined by considering how much of the compound reacted with one equivalent of an acid like HCl. Using current terminology, one mole of $Fe(NO_3)_3$ would react with three moles of HCl and so the equivalent weight of $Fe(NO_3)_3$ would be one-third of the molecular weight or 80.6 g. In the case of Fe_2O_3, one mole reacts with six moles of HCl and so the equivalent weight of Fe_2O_3 is one-sixth of the molecular weight or 26.6 g.

Gladstone's [4] comment that one equivalent of Fe_2Rd_3 would react with three equivalents of MS_2Cy is misleading from the point of view of the theory of equivalents. If 'x' grams of A reacted with 'y' grams of B then 'x' grams of A was regarded as having the same number of equivalents as 'y' grams of B. Using current terminology and reaction (3.3) the number of equivalents in one mole of $Fe(NO_3)_3$ would be 241.85 g divided by one-third the molar mass (80.6 g) giving three equivalents. Since the equivalent weight and the molecular weight are the same for KSCN, then the number of equivalents in three moles of KSCN would also be three equivalents. It would appear that Gladstone was using the word 'equivalent' rather loosely by referring to the numerical coefficients preceding the formulae in Eq. (3.1).

In the middle of the 19th century in Britain, masses were measured in 'grain measures' with 1 grain measure being equal to 64.8 mg. Gladstone [9] speaks about preparing a solution of ferric chloride "containing an amount of iron equivalent to 0.162 grain measures of sesquioxide [Fe_2O_3 in our current format] in every 1000 grain measures of water." Nominally speaking, a grain measure referred to the mass of a grain of barley but grains of barley differed considerably in mass and so some standard needed to be adopted. The standard referred to the mass of 251.428 equivalent pieces of brass that could balance the mass of a cubic inch of water at a pressure of 30 in. of mercury and 62 °F [15]. It is these details that led to the value of 64.8 mg for a grain measure. Using current terminology, the number of moles of iron in 0.162 grain measures of Fe_2O_3 is 1.31×10^{-4} mol and, according to Bergman's view of complete reaction, this was expected to react completely with 3.93×10^{-4} mol of the sulphocyanide of potassium.

The British parliament adopted the platinum standard pound as the standard unit of mass in 1855 with the Weight and Measurement Act of 1878 prohibiting the use of the French metric system of grams and kilograms. Within the scientific community, however, the need to standardise the unit of mass internationally ultimately led to the adoption of the standard kilogram as a mass standard. It is the mass of a platinum-iridium cylinder originally stored in a vault at the International Bureau of Weights and Measures in France in 1901. This represented a major difference in standards from a brass piece representing a grain of barley to a cylinder of a platinum alloy. The name, 'SI units' (*Système International d'Unités*) was adopted in 1960. In May 2019, the standard kilogram was redefined by the *International Commission for Weights and Measures* in terms of Planck's constant, measured on a kibble balance to be $6.62607015 \times 10^{-34}$ kg m^2 s^{-1}.

The quantitative approach to chemical reactions involving the use of equivalent weights or atomic weights and standards by which to measure the mass of a substance demanded of the chemist the use of pure compounds and the achievement of this could present difficulties for the chemist. Naturally occurring iron compounds would most likely coexist with other metal compounds such as those of manganese for example. The techniques for purification were an important skill to have for the 19th century chemist. Gladstone [9] refers to "a solution of pure sulphocyanide of potassium" being prepared and refers to the difficulty of preparing pure iron salts: "The salts were made principally by dissolving pure hydrated ferric oxide in the pure acid; but it was found very difficult to obtain them of a definite constitution. Yet those employed, if not absolutely coinciding with the expression Fe_2Rd_3 came very close to it…".

Much of Gladstone's paper was dedicated towards attempting to resolve a controversy related to the very nature of a chemical reaction and some detail of this controversy is given in the next chapter.

References

1. Leicester HM (1971) The historical background of chemistry. Dover Publications, New York, p 145
2. Crosland MP (1962) Historical studies in the language of chemistry. Heinemann, London, p 185
3. Chang H (2012) Is water H_2O? Evidence, realism and pluralism. Springer, Dordrecht
4. Gladstone JH (1855) On circumstances modifying the action of chemical affinity. Phil Trans R Soc Lond 145:179–223
5. Dalton J (1808) A new system of chemical philosophy. R. Bickerstaff, Manchester
6. Deville MH (1849) Note sur la production de l'acide nitrique anhydre. Comptes Rendus 28:257–260
7. Crosland MP (1962) Historical studies in the language of chemistry. Heinemann, London Chap. 7
8. Ihde AJ (1964) The development of modern chemistry. Harper and Row, New York, p 185
9. Gladstone JH (1855) On circumstances modifying the action of chemical affinity. Phil Trans R Soc Lond 145:182
10. Stokes HN, Cain JR (1907) On the colorimetric determination of iron with special reference to chemical reagents. J Am Chem Soc xxix(4):409–447
11. Philip JC, Bramley A (1913) The reaction between ferric salts and thiocyanates. J Chem Soc Trans 103:795–807
12. Marshall H (1902) Suggested modifications of the sign of equality for use in chemical notation. Proc Edin Roy Soc 24:85–87
13. Bailey KC (1924) The reaction between ferric chloride and potassium thiocyanate. Proc Roy Irish Acad B 37:6–15
14. Kielland J (1937) Individual activity coefficients of ions in aqueous solution. J Am Chem Soc 59(9):1675–1678
15. Alexander JH (1857) Universal dictionary of weights and measures. Wm.Minifie, Baltimore

Chapter 4
The Reaction: Chemical Affinity and Controversy

4.1 Gladstone and 'Complete' or 'Incomplete Reaction'

In 1855, the nature of a chemical reaction was still being hotly debated amongst chemists. They drew upon the concept of 'chemical affinity' to explain why substance A might react with substance B, but not substance C, in that A had a stronger affinity for B, than it did for C, or was more strongly attracted to B than to C. Since the 18th century, attempts had been made to construct affinity tables for certain classes of reaction, such as reactions of elements and compounds with sulfuric acid, to take just one example. Substances high in the table would show a stronger affinity for sulfuric acid than substances lower in the table [1]. Towards the end of the 18th century there were two different views emerging as to the impact of affinity on a chemical reaction. The Swedish chemist, Torbern Olof Bergman (1735–1784), suggested, after a study of many reactions, that if A reacted with B, it did so completely. The French chemist, Claude Louis Berthollet (1748–1822), believed that A need not react completely with B. Many of the reactions studied by Bergman were precipitation reactions and Berthollet suspected that the reaction was due to insolubility factors rather than affinity. In examining these two differing views of reaction, Gladstone [2] (Table 1.1), who would eventually succeed Faraday as Fullerian professor of chemistry at the Royal Institution, chose to study the iron(III) thiocyanate reaction because it did not involve precipitation. Furthermore, its colour was unique amongst iron compounds so there was no mistaking the fact that the deep red colour indicated the presence of the thiocyanate of iron and not some other compound of iron.

Gladstone [3] sets out the Bergman and Berthollet proposals as follows:

A mixture of two salts in solution, which do not produce a precipitate, affords a case where this requisite is fulfilled. Let AB and CD be such salts. According to the one view, when mixed they will either remain without mutual action, or, should the affinities so preponderate, they will become simply AC and BD, the excess of either original salt remaining inactive. According to the other view, A will divide itself in certain proportions between B and D, while C will do the same in the inverse ratio, the said proportions being determined not solely by the differences of energy in the affinities, but also by the differences of the quantities of the bodies.

© The Author(s), under exclusive license to Springer Nature Switzerland AG 2019
K. C. de Berg, *The Iron(III) Thiocyanate Reaction*, SpringerBriefs in
History of Chemistry, https://doi.org/10.1007/978-3-030-27316-3_4

Gladstone [4] set about comparing the colours of iron(III) thiocyanate mixtures in clear glass vessels, relying on his assistant to make the final judgment: "My own observation was always checked by that of my assistant, and if we differed I generally adopted his view, since having no idea of what result was to be expected, his judgment was more impartial." The test with iron(III) thiocyanate is described as follows [4]:

> The first object to be determined evidently was, whether on mixing three equivalents of sulphocyanide of potassium with one equivalent of the ferric salt, say the chloride, the full depth of colour possible from the combination of all the sulphocyanagen with all the iron was actually obtained. That this was not the case was seen at once, for on the addition to such a mixture of either more sulphocyanide of potassium, or more chloride of iron, the colour was increased.

Thus Gladstone resolved the situation in favour, in his mind, of Berthollet's view, which turned out to be an important criteria for the concept of chemical equilibrium. But is '*incompleteness of reaction*' the only way to interpret Gladstone's experimental observations?

4.2 Identity of Species Responsible for the Blood-Red Colour

A scan over Table 1.1 reveals that there was also considerable debate about the identity of the chemical species responsible for the red colour on mixing ferric ions with thiocyanate ions. Over 160 years, candidates included $Fe(SCN)_3$, $Fe\{Fe(SCN)_6\}$, $FeHC_3N_3S_3O_3$, $Fe(SCN)_6^{3-}$, $Fe(SCN)^{2+}$, $Fe(SCN)^+$, and $Fe(SCN)_2^+$. It is understandable why one of the first candidates was $Fe(SCN)_3$ because this was expected from the stoichiometry of the reaction. The application of Job's method of continuous variations led eventually to the identification of $Fe(SCN)^{2+}$, and $Fe(SCN)_2^+$ as the dominant species up to thiocyanate concentrations of 0.25 M in aqueous solution and most likely $Fe(SCN)_3$ for the red colour in organic extracts [5–7].

Hill and MacCarthy [8] give a clear explanation of Job's method. For a metal complexation reaction of the general type: $M + nL \rightleftharpoons ML_n$; equimolar solutions of metal and ligand are mixed in different metal:ligand ratios, $(1 - X)$ mole fraction of M with X mole fraction of L, in such a way that the metal concentration plus the ligand concentration is constant for each mixture. The absorbance of each mixture, corrected for metal and ligand absorbance in the absence of complexation, is plotted against the mole fraction of metal or ligand. The mole fraction at the maximum absorbance leads to the composition of the complex ML_n where, "$n = X/(1 - X)$" . Thus, a maximum at X (mole fraction) equal to 0.5 indicated a 1:1 metal:ligand complex; a maximum at X equal to 0.67 indicates a 1:2 metal:ligand complex; and a maximum at X equal 0.75 indicates a 1:3 metal:ligand complex. This method was used as early as 1942 to determine composition according to Table 1.1. Hill and MacCarthy used a 0.001 M solution of $Fe(NO_3)_3$ and 0.001 M solution of KSCN to make up the different composition mixtures as stipulated above and plotted absorbance against

Fig. 4.1 The plot of absorbance, A, against mole fraction, X_{SCN^-}, for mixtures prepared from 0.001 M solutions of $Fe(NO_3)_3$ and KSCN

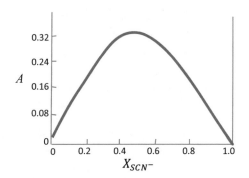

mole fraction of thiocyanate as shown in Fig. 4.1. One can see that the mole fraction at the maximum is 0.5 giving a value of 'n' equal to one corresponding to the species, $FeSCN^{2+}$, or, showing coordinated water, $Fe(H_2O)_5SCN^{2+}$.

As recently as 2011, some chemists [9] have claimed that all iron:thiocyanate species from 1:1 through to 1:6 can be identified in solutions up to about 6.5 M thiocyanate but there is some doubt as to whether the instability of the complexes was fully accounted for. The instability of the complexes evidenced by the fading of the red colour over time was noted as early as 1907 (Table 1.1), and probably before this. Sultan and Bishop [10] could only justify the presence of the 1:1 complex even when the SCN^- concentration was 5000 times the Fe^{3+} concentration. Najib and Hayder attribute the instability to an inner redox reaction between SCN^- and CNS^-. This is the only paper in Table 1.1 that gives this reason for instability. The most common reasons given for the instability have been the reduction of iron(III) to iron(II) by SCN^- and exposure of the system to light. Further work needs to be done on this problem. Chemists have tried to avoid the problem by conducting experiments in the dark or by adding an oxidising agent like hydrogen peroxide to the mixture. It would appear that the most efficient way to deal with the problem, given some confusion over the chemical reason for the instability, is to determine the initial absorbance spectrum at time zero before instability can impact the results [7].

The foregoing serves to illustrate that controversy is endemic to chemistry and it is as controversies are resolved that knowledge is gained. A similar process of resolution of controversy occurs in most fields of scientific endeavour. The public, and often our chemistry students, expect science to proceed without controversy. But, as chemists, it can take some time before tools to access the invisible world of atoms and molecules, like Job's method for example, become available. And even the presence of these tools is not guaranteed to eliminate controversy because tools may have limited applicability. There is still some controversy associated with the iron(III) thiocyanate reaction we have known about for nearly 200 years.

4.3   Can One Verify 'Incompleteness' for the Iron(III) Thiocyanate Reaction?

Fourteen of the papers presented in Table 1.1 were published in the *Journal of Chemical Education* demonstrating the interest shown by chemistry educators across high school, college and university chemistry in the iron(III) thiocyanate system. One such paper is that by Ghirardi et al. [11] which examined a teaching sequence for learning the concept of chemical equilibrium. Fourth in the sequence was the iron(III) thiocyanate equilibrium. The point of using this equilibrium was to demonstrate that equilibria are *incomplete transformations* on the basis of studying what happens when crystals of $Fe(NO_3)_3$ and KSCN are separately added to the equilibrium. And it is this study which gives us a clue to answering the question previously posed: Is *'incompleteness of reaction'* the only way to interpret Gladstone's experimental observations? What is interesting is that some students had great difficulty in accepting the idea of incomplete transformation, even though the experiment was basically a repeat of the 1855 study by Gladstone outlined above. This confirms the idea that there is no guarantee that our current students will come to the same conclusion as the professional chemists of the time when studying a reaction of historical significance. The only way the instructor could get students to *reluctantly* accept "*incomplete transformation*" was to take it as a *working hypothesis* and ask, "Does the data support this hypothesis?"

It is interesting to ponder whether Gladstone's experiment of adding additional iron(III) and additional thiocyanate to the iron(III) thiocyanate mixture could also have been interpreted in terms of Bergman's "*complete transformation*" on the basis of what we now know about the reaction. The reason why some students in the Ghirardi et al. study found it difficult to accept the model of "*incomplete transformation*" was that their chemistry education to this point had been based on the "*limiting reagent*" concept in chemical reactions, which depended on the idea of "*complete transformation*" of at least one of the reactants. The research results in Table 1.1, particularly the possible identification of the species responsible for the red colour previously summarised, can assist in answering this question. A comparison of the Berthollet and possible Bergman explanations and assumptions for the Gladstone experiment is shown in Table 4.1. The Bergman explanations have been based on what some of the students in the Ghirardi et al. study suggested [11]. These students explained the increase in intensity of the blood-red colour as due to simply adding more iron or more thiocyanate to $FeSCN^{2+}$ rather than there being residual Fe^{3+} to react with added SCN^- or residual SCN^- to react with added Fe^{3+}. Whether the students were thinking of species like Fe_2SCN^{5+} for the addition of more iron and $Fe(SCN)_2^+$ for the addition of more thiocyanate is hard to tell, but research has certainly confirmed the existence of $Fe(SCN)_2^+$ while $Fe\{Fe(SCN)_6\}$, a species containing more iron, was suggested at one point in the history of the reaction research according to Table 1.1.

The analysis shown in Table 4.1 is designed to show that one could have interpreted Gladstone's observations in terms of *'complete transformation'*, for

Table 4.1 Explanations for the increase or decrease in intensity of the blood-red colour of $FeSCN^{2+}$ when solid $Fe(NO_3)_3$, KSCN, and aqueous $HgCl_2$ are separately added to, $Fe^{3+} + SCN^- \rightarrow FeSCN^{2+}$, based on incomplete transformation (Berthollet) and complete transformation (Bergman)

Action	Observation	Explanation	Assumptions
Addition of $Fe(NO_3)_3(s)$	Blood-red colour intensifies	*Bergman:* $$FeSCN^{2+} + Fe^{3+} \rightarrow Fe_2SCN^{5+}$$ *Berthollet:* $$Fe^{3+} + SCN^- \rightarrow FeSCN^{2+}$$	*Bergman:* Fe_2SCN^{5+} exists and has a blood-red colour *Berthollet:* Residual SCN^- exists to react with added Fe^{3+}
Addition of KSCN(s)	Blood-red colour intensifies	*Bergman:* $$FeSCN^{2+} + SCN^- \rightarrow Fe(SCN)_2{}^+$$ *Berthollet:* $$Fe^{3+} + SCN^- \rightarrow FeSCN^{2+}$$	*Bergman:* $Fe(SCN)_2{}^+$ exists and has a blood-red colour *Berthollet:* Residual Fe^{3+} exists to react with added SCN^-
Addition of $HgCl_2(aq)$	Blood-red colour loses intensity	*Bergman:* $$FeSCN^{2+} + HgCl_2 \rightarrow FeCl_2{}^+ + HgSCN^+$$ *Berthollet:* $$Hg^{2+} + SCN^- \rightarrow HgSCN^+$$ $$FeSCN^{2+} \rightarrow Fe^{3+} + SCN^-$$	*Bergman:* $HgSCN^+$ and $FeCl_2{}^+$ exist and do not have a blood-red colour *Berthollet:* $HgSCN^+$ exists and does not have a blood-red colour. The iron(III) thiocyanate reaction is reversible

example, $FeSCN^{2+} + SCN^- \rightarrow Fe(SCN)_2{}^+$, rather than Fe^{3+}(remaining) + SCN^-(additional) $\rightarrow FeSCN^{2+}$(more). The 'Assumptions' column in Table 4.1 is a clue as to what questions could be asked to try to resolve the controversy and suggested questions are shown in Table 4.2. The questions on the Bergman side are quite feasible given that we now know of the existence of $Fe(SCN)_2{}^+$. The fact that $Fe(SCN)_2{}^+$ absorbs at a wavelength only 20 nm higher than that for $Fe(SCN)^{2+}$ and the fact that $Fe(SCN)_2{}^+$ has a molar absorptivity about double that for $Fe(SCN)^{2+}$

Table 4.2 Some important questions arising from the Bergman and Berthollet explanations in Table 4.1

Bergman questions	Berthollet questions
1. Is there evidence that Fe_2SCN^{5+} exists and has a colour similar to blood-red?	1. In an equimolar reaction between Fe^{3+} and SCN^-, is there evidence of unreacted Fe^{3+} and SCN^- when the reaction has finished?
2. Is there evidence that $Fe(SCN)_2{}^+$ exists and has a colour similar to blood-red?	2. Is the iron(III) thiocyanate reaction reversible?
3. Is there evidence that $HgSCN^+$ and $FeCl_2{}^+$ exist and do not have a blood-red colour?	3. Is there evidence that $HgSCN^+$ exists and does not have a blood-red colour?

means the colour of the solution is still basically blood-red visually but more intense when $Fe(SCN)_2^+$ forms.

It turns out to be very difficult to answer Question 1 on the Berthollet side because reactants cannot be separated from the products in the aqueous solution reaction to test for the presence of unreacted Fe^{3+} and SCN^-. For precipitation equilibria like: $AgCl(s) \rightleftharpoons Ag^+(aq) + Cl^-(aq)$, the system can be filtered to separate $AgCl(s)$ from the other components. Addition of $KI(aq)$ to the filtrate will precipitate yellow $AgI(s)$ showing that residual $Ag^+(aq)$ must have been present in the filtrate. Reversibility is also more easily demonstrated with precipitation equilibria using radioactive isotopes. Adding labelled silver nitrate to the silver chloride equilibrium, for example, leads to the presence of radioactivity in the silver chloride precipitate demonstrating reversibility and a dynamic equilibrium. The Berthollet explanations also illustrate the importance of *quantity* in determining *affinity* as Gladstone affirmed. Adding excess ferric ion increased its affinity with the remaining thiocyanate ion to produce an enhanced amount of the red coloured species. On the other hand, Bergman explained the enhanced red colour in terms of an ongoing reaction rather than increased affinity of ferric ion for remaining thiocyanate ions. So deciding between the Bergman and Berthollet explanations is not a straightforward matter in the case of the iron(III) thiocyanate reaction. While Gladstone interpreted the changes in Table 4.1 in terms of the Berthollet explanation of incomplete transformation, some students had more difficulty in accepting the Berthollet explanation as evidenced by the Ghirardi et al. study. This is a case where two different interpretations of the same empirical evidence seems to apply.

Gladstone [12] also observed that, "A solution of chloride of mercury …very speedily removes the colour", and this is presented as the third item in Tables 4.1 and 4.2. The species, $HgSCN^+$, is a coordination compound on the way to producing the species $Hg(SCN)_2$ which is not very soluble in water. Both the Bergman and Berthollet explanations in Tables 4.1 and 4.2 rely on the production of this mercuric compound although other possibilities such as a combined coordination compound between $FeSCN^{2+}$ and Hg^{2+} or Cl^- could be posited but we have selected the simplest option for illustration purposes. For the first two cases in Tables 4.1 and 4.2 the Berthollet explanation does not depend on the formation of new compounds as required in the Bergman explanation but deciding between the two explanations can still prove difficult particularly for students. Berthollet and Bergman were both highly respected chemists of their day but differed in their understanding of a chemical reaction. So students should not be surprised if it is difficult to choose between the two approaches. If students realize that this situation is endemic to how knowledge is often generated in chemistry and they learn the art of asking the appropriate questions, that is, learn to interrogate chemical ideas, much will have been achieved in their chemistry education. It took studies of a large number of reactions to reach the conclusion that equilibrium reactions were incomplete transformations, reversible and dynamic and that the chemical affinity between two species depended not only on their innate or elective affinity as Bergman called it but also on the quantity of the species involved. So it is misleading to think that all the properties of chemical equilibria can be discovered by looking at just one reaction like the iron(III) thio-cyanate reaction. It is probably more appropriate to think of the iron(III) thiocyanate reaction as a means of illustrating the principles of chemical equilibria rather than discovering those principles.

4.4 Berthollet and Affinity

About 50 years before Gladstone's (1855) experiments with the iron(III) thiocyanate reaction Berthollet had become convinced that quantity of species was an important contributor to chemical affinity. Holmes [13] considers that this conviction probably arose from the fact that he was looking at chemical reactions on a grand scale,

> which he gained while directing the efforts of the French Revolutionary government to establish new chemical industries. He discovered, for example, that during the extraction of saltpetre from crude nitre rock by dissolution in water, the increasing concentration of saltpetre in solution made the remaining portions more difficult to dissolve, even though the water was never saturated with salt.

Saltpeter (potassium nitrate) is used in explosives so it is no wonder that the French government was interested in its acquisition. Berthollet interpreted this phenomenon as the result of an increase in the affinity of the species in solution thus increasing the tendency of the reverse reaction to produce solid saltpetre. In modern terms, the dissolution would be represented as: $KNO_3(s) \rightleftharpoons K^+(aq) + NO_3^-(aq)$. While accompanying Napoleon's expedition to Egypt in 1798, Berthollet observed large amounts of sodium carbonate being produced from salt water lakes on the edge of the desert due to the reaction: $2NaCl + CaCO_3 \rightarrow Na_2CO_3 + CaCl_2$, and the continual removal of sodium carbonate as it formed. This reaction is quite unsuccessful in producing sodium carbonate under laboratory conditions, but in circumstances where large quantities of saltwater and limestone are present, copious amounts can be produced. Holmes [14] explains this phenomenon as follows: "The large quantities in solution of the original substances compared to those of the products maintained a reaction which would not take place by affinities alone." The reference to affinities in this statement is to '*elective affinities*' which by definition were independent of quantity according to Bergman. Berthollet referred to his new concept of affinity as *chemical mass* which encompassed the notion of elective affinity in addition to the contribution of quantity. It must be remembered that the term 'mass' here is different from our current understanding of the term.

Berthollet was also convinced of the reversibility of reactions by a consideration of reactions like: $Na_2CO_3 + BaSO_4 \rightleftharpoons Na_2SO_4 + BaCO_3$. By adding excess alkali (Na_2CO_3) to barium sulfate he was able to convert most of the barium sulfate to sodium sulfate. Evaporating the mixture to dryness, removing excess alkali with alcohol, and dissolving the residue in water he was able to convert most of the sodium sulfate back to barium sulfate. Even though barium has a greater affinity for sulfate than does sodium, Berthollet was able to adjust the quantities so that sodium sulfate could form. One of the problems for demonstrating reversibility in the iron(III) thiocyanate reaction is that neither ferric ion or thiosulphate ion are coloured under acidic conditions. So one cannot demonstrate reversibility by looking for colour changes. In a reaction like:

$$Cr_2O_7^{2-}(aq) + H_2O(l) \rightleftharpoons 2CrO_4^{2-}(aq) + 2H^+(aq),$$

where $Cr_2O_7^{2-}(aq)$ is orange and $CrO_4^{2-}(aq)$ is yellow, the addition of base will force the forward reaction to occur with a colour change from orange to yellow, and the addition of acid will force the reverse reaction to occur with a colour change from yellow to orange. These observations help to confirm the reversibility of the reaction. This was not possible in the iron(III) thiocyanate reaction.

Fowles' 1937 lecture experiments in chemistry [15] considers the iron(III) thiocyanate reaction equilibrium as:

$$FeCl_3 + 3NH_4CNS \rightleftharpoons Fe(CNS)_3 + 3NH_4Cl$$

The stated purpose of the lecture demonstration was to show, through colour change, how the direction of equilibrium may be altered by adding more $FeCl_3$, more NH_4CNS, and more NH_4Cl. This experiment was not designed to illustrate Le Chatelier's principle, although it could do that, but by showing how the rates of the forward and reverse reactions can be changed. Fowles made no mention of the concept of ionic strength and interpreted the reduced colour on adding NH_4Cl as a product of the increased rate of the reverse reaction. Driscoll [16] took up this same issue and showed that the reduced colour also happens when sodium nitrate is added. If one considers the reaction in the form: $Fe^{3+} + SCN^- \rightleftharpoons FeSCN^{2+}$, one can see that adding solid NH_4Cl or $NaNO_3$ doesn't change the concentration of either of the three species in the ionic equation. The addition of these salts does change the ionic strength and ultimately the activity of the ions in the equation and the reaction will move in a direction that opposes the change in activity. The concept of *activity* takes into account the electrical interactions between ions in solution and consequently measures 'effective concentration' in the light of these interactions. This concept will be considered in further detail in the next chapter.

References

1. Leicester HM (1971) The historical background of chemistry (Dover ed). Dover Pub Inc., New York, pp 125–129. (Originally published by Wiley, New York, 1956)
2. Gladstone JH (1855) On circumstances modifying the action of chemical affinity. Phil Trans R Soc Lond 145:179–223
3. Gladstone JH (1855) On circumstances modifying the action of chemical affinity. Phil Trans R Soc Lond 145:181
4. Gladstone JH (1855) On circumstances modifying the action of chemical affinity. Phil Trans R Soc Lond 145:183

5. Perrin DD (1958) The ion $Fe(CNS)_2^+$. Its association constant and absorption spectrum. J Am Chem Soc 80(15):3852–3856

6. Bjerrum J (1985) The iron(III)-thiocyanate system. The stepwise equilibria studied by measurements of the distribution of tris(thiocyanato)iron(III) between octan-2-ol and aqueous thiocyanate solutions. Acta Chem Scand A 39:327–340

7. de Berg K, Maeder M, Clifford S (2016) A new approach to the equilibrium study of iron(III) thiocyanates which accounts for the kinetic instability of the complexes particularly observable under high thiocyanate concentrations. Inorg Chim Acta 445:155–159

8. Hill ZD, MacCarthy P (1986) Novel approach to Job's method. J Chem Educ 63(2):162–167

9. Najib FM, Hayder OI (2011) Study of stoichiometry of ferric thiocyanate complex for analytical purposes including F^- determination. Iraqi Nat J Chem 42:135–155

10. Sultan SM, Bishop E (1982) A study of the formation and stability of the iron(III)-thiocyanate complex in acidic media. Analyst 107:1060–1064

11. Ghirardi M, Marchetti F, Pettinari C, Regis A, Roletto E (2014) A teaching learning sequence for learning the concept of chemical equilibrium in secondary school education. J Chem Educ 91(1):59–65

12. Gladstone JH (1855) On circumstances modifying the action of chemical affinity. Phil Trans R Soc Lond 145:185

13. Holmes FL (1962) From elective affinities to chemical equilibria: Berthollet's law of mass action. Chymia 8:105–145, 109

14. Holmes FL (1962) From elective affinities to chemical equilibria: Berthollet's law of mass action. Chymia 8:105–145, 110

15. Fowles G (1937) Lecture experiments in chemistry. G. Bell & Sons Ltd., London

16. Driscoll DR (1979) Invitation to enquiry: the Fe^{3+}/CNS^- equilibrium. J Chem Educ 56(9):603

Chapter 5
The Reaction: Chemical Affinity: Laws, Theories and Models

5.1 Geoffroy's Table of Affinity 1718

According to Quílez [1] the idea of *affinity* as the disposition of two chemical species to react goes back to the work of the theologian and natural philosopher Albertus Magnus during the 13th century. The concept held theoretical status for hundreds of years with scholars having difficulty knowing how to define it quantitatively. Knight [2] pictures Lavoisier as failing to do this even into the 18th century: "Because Lavoisier refused…to deviate from the law of forming no conclusions which are not fully warranted by experiment, he had not felt able to tackle the theory of affinity." This seems a difficult argument to sustain given the large number of experiments which would have had to be conducted to draw up an affinity table like that presented by Geoffroy in 1718 [3], well before the time of Lavoisier, and shown in Fig. 5.1. However, Lavoisier would not have argued with the experimental results, but only with invoking the theory of affinity to explain them. It took a long time for chemists to understand the delicate relationship that exists between experiment and theory: Lavoisier could not interpret his experiments without recourse to caloric theory, but did not always recognise this; Priestley could not interpret his experiments without reference to phlogiston theory; Newton could not interpret his law of falling bodies without a theory of gravity; and Geoffroy, Bergman, and Berthollet could not interpret chemical reactions without reference to the theory of affinity. Theory was often equated with speculation and hypotheses, activities that should be avoided by chemists at all costs according to Priestley and Lavoisier.

Lavoisier's and Priestley's subtle reliance on theory in spite of their declaration to always give precedence to observation and experiment is discussed by de Berg [4]. While no argument could be advanced against Newton's law of falling bodies, his reference to gravity was severely criticised [5]. Probably due to this criticism, chemists tried to avoid likening affinity in chemical reactions to gravitational attraction in physics.

How was a chemist meant to interpret the data in Fig. 5.1? Consider the third column as an example. At the head of the column is the symbol for nitric acid (or

K. C. de Berg, *The Iron(III) Thiocyanate Reaction*, SpringerBriefs in
History of Chemistry, https://doi.org/10.1007/978-3-030-27316-3_5

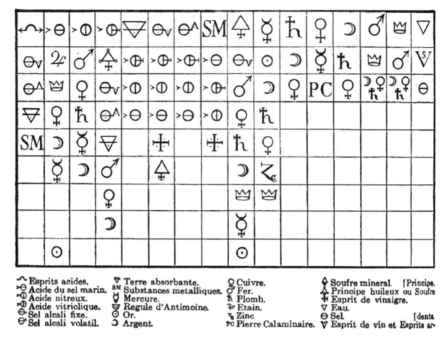

Esprits acides.   Terre absorbante.   Cuivre.   Soufre mineral. [Principe.
Acide du sel marin.   Substances metalliques.   Fer.   Principe huileux ou Soufre
Acide nitreux.   Mercure.   Plomb.   Esprit de vinaigre.
Acide vitriolique.   Regule d'Antimoine.   Etain.   Eau.
Sel alcali fixe.   Or.   Zinc   Sel. [denta
Sel alcali volatil.   Argent.   Pierre Calaminaire.   Esprit de vin et Esprits ar

Fig. 5.1 An affinity table constructed by Geoffroy in 1718 [3]

often referred to as nitrous acid in the 18th century) followed from top to bottom by the symbols for the metals Fe, Cu, Pb, Hg, and Ag. The column summarises metal displacement reactions in nitric acid so a metal added to the nitrate of a metal below it in the column should displace that metal from its nitrate. So Pb should be able to displace Ag from a solution of $AgNO_3$ according to the reaction: $2AgNO_3(aq) + Pb(s) \rightarrow 2Ag(s) + Pb(NO_3)_2$. This reaction was understood to mean that lead, Pb, has a stronger affinity for nitric acid than does silver, Ag. So iron, Fe, has a stronger affinity for nitric acid than does copper, Cu, and one would expect the reaction; $Cu(NO_3)_2(aq) + Fe(s) \rightarrow Cu(s) + Fe(NO_3)_2(aq)$, to occur. Thus Geoffroy's table of affinities became a useful practical and organisational tool of chemical reactions for chemists. What was once considered a theory of affinity took on the properties of a law of affinity at least in Geoffroy's mind [6]: "For the *laws* of these rapports I *have observed* that when two substances having a *disposition* to unite together are combined and a third added, the third may part the two taking one or the other." Just as the law of periodicity can be used to describe the organisation of the periodic table, so the law of affinity can be used to describe the organisation of tables of affinities. While the table of affinities did not quantify the concept of affinity, the fact that relative affinities were represented did give a semi-quantitative flavour to the concept.

5.2   Waage and Guldberg and the Law of Mass Action

While Newton's introduction of the concept of gravity attracted significant criticism, the level of quantification achieved with the help of mathematics in a field like mechanics was duly admired by physicists and chemists alike. The question was: Could *affinity* achieve a level of quantification like that experienced in physics? By the middle of the 19th century we have seen that *incomplete transformation* and *reversibility* had come to characterise the nature of a chemical reaction and the concept of affinity had come to be represented by two characteristics, that of inherent or quantity-independent or what was known as *elective affinity*, a term coined by Bergman, and the quantity-dependent component of affinity called by Berthollet, *chemical mass*. In 1864, the Norwegian chemist Peter Waage (1833–1900) and mathematician Cato Guldberg (1836–1902) put these ideas together to form what became known as the law of mass action or the law of chemical equilibrium.

Interestingly, Waage and Guldberg [7] call upon the work of Gladstone [8] and 'certain colour reactions' like the apparently incomplete iron(III) thiocyanate reaction to support their theory of chemical equilibrium in the following words:

> One has tried even earlier to apply our view [theory]…of the equilibrium state…for mixtures of two different soluble salts from which no precipitation occurs. One has…partly with the help of certain colour reactions, partly with the help of the rotation of the plane of polarization (*Gladstone*) and partly with the help of diffusion experiments (*Graham* and *Gladstone*), sought to demonstrate that a partial substitution of the soluble salts occurs.

Some chemical reactions like precipitation reactions used by Bergman to support his theory of complete reaction seem to proceed in only one direction and Waage and Guldberg [9] give the following explanation of this phenomenon using equilibrium principles based on the concept of *force*:

> If processes where either only one or the other direction is apparent seem to occur often in chemistry, it frequently arises from the combining or acting force being either very large or very small in relation to the opposite one under the conditions present. If one modifies the conditions under which the forces operate in one way or the other, then one will either cause the opposing force to become about as strongly effective as the first—and in such a case both directions of the process will be apparent simultaneously—or one will even be able to cause the opposing force to be dominant, so that only the second part of the process appears to occur.

The *law* of mass action, the ancestor of the *law* of chemical equilibrium, proposed by Waage and Guldberg was based on equating the *action force* of two reactant substances, P and Q, with the *reaction force* of two product substances, Ṕ and Q́. If all masses, known as *active* masses occur in the same volume, then the chemical *force* contributed by each substance is proportional to its active mass raised to a power determined only by experiment. Here the word *force* is being used for *affinity* and as we have seen not all chemists would have been happy with the word *force* given its connotations with Newton's *force* of gravity. Waage and Guldberg differed from Berthollet who considered affinity to be proportional to mass not mass raised to some exponent. A *description* of Waage and Guldberg's [10] understanding of chemical equilibrium for the reaction, $P + Q = \acute{P} + \acute{Q}$, appears in their *law* of mass action as:

$$[alpha](p - x)^a (q - x)^b = [alpha'](p' + x)^{a'} (q' + x)^{b'}$$

Here, the terms $[p, q, p', q']$ refer to initial active masses and the term $[x]$ refers to the mass that has reacted at equilibrium. The powers were later shown to be simply the coefficients in the stoichiometric equation. It must be remembered that the term *mass* has a different connotation to the term as it is used today. As a term for amount or quantity it resembles more our idea of particle number or mole as used in the 20th century. Masses in a unit volume were eventually replaced with concentrations in units of mol L^{-1}, and finally concentrations were replaced with 'effective concentrations' or activities which accounted for ionic interactions in solution. The impact of ionic interactions is expressed through the use of an *activity coefficient*, γ, which normally takes on a value between 0 and 1, the value 1 indicating that ionic interactions are negligible. In solutions, activity 'a' is expressed as,

$$a = \gamma \frac{m}{m^0},$$

where 'm' is the molality and m^0 is the standard molality 1 mol kg^{-1}. In practice molality can be replaced with molarity for dilute solutions. The proportionality coefficients $[alpha]$ and $[alpha']$ are the affinity coefficients. Thus the law of mass action for the production of the 1:1 complex for the reaction: $Fe^{3+}(aq) + SCN^-(aq) = FeSCN^{2+}(aq)$, could be written as:

$$k_f \{Fe^{3+}\}\{SCN^-\} = k_r \{FeSCN^{2+}\}$$

where {species} represent activities and k_f and k_r represent the affinity coefficients for the forward and reverse reactions but which were to become the rate constants reflecting the fact that one now understands equilibrium to represent equality of rates of reaction rather than equality of forces. The law of chemical equilibrium is most commonly now expressed using the ratio of the rate constants, k_f/k_r, taken to be the equilibrium constant, K. So:

$$K = \frac{\{FeSCN^{2+}\}}{\{Fe^{3+}\}.\{SCN^-\}}$$

which is the form of the law used in de Berg et al. [11]. For dilute solutions activities approximate to concentrations, [species], so K is sometimes expressed under these circumstances as,

$$K = \frac{[FeSCN^{2+}]}{[Fe^{3+}].[SCN^-]}.$$

With the expression for K we see the beginnings of a quantitative approach to the old concept of affinity. If Fe^{3+} and SCN^- have a high affinity for each other, one would expect the value of K to be large. Alternatively, if Fe^{3+} and SCN^- have a low affinity for each other, one would expect the value of K to be small.

5.3 Law, Theory and Model

The terms 'law', 'theory', and 'model' are often used interchangeably in common *chemistry speak*. However, philosophers claim that much is to be gained by trying to distinguish between these terms [12–14]. This proves to be helpful when it comes to understanding the historical significance of the iron(III) thiocyanate reaction if one thinks of *law* as a *description*, *theory* as an *explanation*, and *model* as a *representation*. There is a tendency to think of laws as a more reliable form of knowledge than theories but the history of the iron(III) thiocyanate reaction shows that laws cannot be separated from their underlying theory. Laws are just as tentative as is the underlying theory and chemical knowledge grows out of both laws and theory [15]. For the reaction in question here, *law* manifests itself as a mathematical description for the equilibrium constant, *model* manifests itself as a chemical equation representation for the reaction, and *theory* manifests itself in terms of an explanation of equilibrium in terms of reaction rates and their equivalence for the forward and reverse reactions.

Applying the law of chemical equilibrium to the data presented by Gladstone [8] and Ghirardi et al. [16] on what happens to the iron(III) thiocyanate equilibrium when solid ferric nitrate, potassium thiocyanate, and aqueous mercuric chloride are added separately to the equilibrium system gives one access to mathematical explanations for the situation observed and discussed in Chap. 4. So while the function of chemical laws is said to 'describe' a chemical situation, explanations can make use of chemical laws just like they can make use of chemical equation models as shown in Tables 4.1 and 4.2 of Chap. 4. The reaction quotient, Q, has the same expression as K above but with concentrations or activities not necessarily equilibrium values. When solid $Fe(NO_3)_3$ is added to the equilibrium system, $[Fe^{3+}]$ increases which means $Q < K$. This means the numerator, $[FeSCN^{2+}]$, must increase to restore the value of K. This can only happen if the forward reaction becomes momentarily favoured over the reverse reaction until $Q = K$. The same reasoning applies to the addition of solid KSCN. In both cases the intensity of the blood-red colour increases which is consistent with the forward reaction being favoured. The addition of $HgCl_2$ reduces the $[SCN^-]$ making $Q > K$, so the reverse reaction becomes momentarily favoured over the forward reaction until $Q = K$. As a result the intensity of the blood-red colour decreases significantly. The qualitative equivalent of this quantitative approach has been called Le Chatelier's principle.

Chemical equations are used to represent equilibria and constitute one way of modelling an equilibrium system. The chemical equation for which the previously written K expression applies is:

$$Fe^{3+}(aq) + SCN^-(aq) \rightleftharpoons FeSCN^{2+}(aq)$$

The equal size half-headed arrows indicate that the equilibrium is a dynamic one in which both the incomplete forward and reverse reactions still occur at equilibrium but at equal rates. The reaction sometimes appears with an equals sign instead of the double-headed arrows as in Driscoll [17] in Table 1.1.

$$FeCl_3 + 3NH_4CNS = Fe(CNS)_3 + 3NH_4Cl$$

Multiple meanings can be ascribed to the equal sign. The sign could indicate conservation of mass, conservation of atoms, conservation of charge, equal rates and so on but the dynamic character is probably best represented by arrows. So the concept of chemical equilibrium was able to explain the observations of Bergman's seemingly complete reactions and those of Berthollet and Gladstone where reactions like the iron(III) thiocyanate reaction did not go to completion. Lund [18] provides a useful analysis of the contribution of Guldberg and Waage to our understanding of equilibrium.

5.4   The Role of Thermodynamics

While a controversy raged between Bergman and Berthollet on the nature of a chemical reaction, both contributed to our understanding of what one might call the driving force of chemical change. Bergman considered 'elective affinities' independent of the quantity of species involved to be the driving force and Berthollet considered the quantity of species (mass) to contribute to the driving force of a chemical reaction. The concept of 'activity' or 'effective concentration' was developed by Lewis and Randall [19] and, according to Lindauer [20], "activity played a role analogous to that of Berthollet's mass", or we could add, to Waage and Gulberg's active mass. Lindauer [21] also adds that, "Wendell M. Latimer's book, "Oxidation Potentials", is the 20th century equivalent of Bergman's 18th century classic on elective affinities." To throw these ideas into perspective it is important to realise that it is from the field of thermodynamics that chemical affinity was eventually quantified. Van't Hoff quantified chemical affinity in terms of the maximum useful work that can be accomplished by the reaction and this became better known as the free energy change associated with the reaction [22]. The classic driving force equation is: $\Delta G = \Delta G^\circ + RT \ln Q$, where ΔG is the free energy change associated with a reaction, $aA + bB \rightarrow cC + dD$; ΔG^0 is the standard free energy change where all activities are taken as 1; R is the gas constant, $8.314\,J\,K^{-1}\,mol^{-1}$; T is the absolute temperature; and Q is the reaction quotient,

$$Q = \frac{\{C\}^c.\{D\}^d}{\{A\}^a.\{B\}^b}.$$

Table 5.1 Highlights in the development of thermodynamics [23]

Author	Date	Contribution
Black	1760–1766	Calorimetry
Mayer	1842	Interconversion of heat and work
Joule	1843–1852	Interconversion of heat and work
Kelvin	1848–1849	Absolute temperature scale
Clausius	1850	Second law
Kelvin	1851	Second law; dissipation of energy
Clausius	1854–1865	Concept of entropy
Gibbs	1873	Chemical thermodynamics
Gibbs	1870–1878	Chemical potential; phase rule
Helmholtz	1882	Theory of equilibrium; free and bound energy
van't Hoff	1884–1887	Chemical thermodynamics; theory of equilibrium constant; solutions
Nernst	1906	Heat theorem (third law)
Simon	1927	Improved version of third law

If A and B have more free energy than C and D, that is, A and B have a stronger affinity for each other than C and D, ΔG will be negative and the forward reaction can do useful work. If C and D have more free energy than A and B, ΔG will be positive, and mixing A and B together will not result in useful work. It is worth reviewing how the driving force equation was developed in the 19th century because it shows how chemistry began to adopt that mathematical approach which proved so effective in physics in the 17th and 18th centuries.

A summary of the highlights in the development of thermodynamics is given in Table 5.1 after Laidler [23]. The developments up to and including Clausius were equally adopted by physics but it was the contributions by Josiah Willard Gibbs (1839–1903), Hermann von Helmholtz (1821–1894), and Jacobus van't Hoff (1852–1911) towards understanding chemical reactions that originated the field of *chemical thermodynamics*. This new field drew upon earlier developments such as an understanding of heat, 'q', and work, 'w', as forms of energy or as processes that lead to changes in the internal energy, U, of a system, $\Delta U = q + w$, the Kelvin scale of temperature T (°C + 273.15), the enthalpy, H, of a system ($H = U + PV$), and the concept of entropy, S, ($\Delta S = q/T$). It was in the field of *chemical thermodynamics* where Gibbs free energy, G, or Helmholtz free energy, A, was introduced as the energy available for doing useful work. The relationship between the state functions G, H and S is, $G = H - TS$, where the analogy that H represents gross income, TS represents rates and taxes, and G represents free spending money is often used. The same analogy can be used with, $A = U - TS$. The derivation of the driving force equation, $\Delta G = \Delta G° + RT \ln Q$, is shown in Box 5.2 after an introduction to the basic relevant principles of calculus is given in Box 5.1 for the benefit of the reader. The derivation in Box 5.2 is designed to draw attention to the fact that infinitesimal calculus, once the sole province of physics and mathematics, became an important part of chemistry and this trend was to continue into the 20th century when quantum mechanics was to service both physics and chemistry in remarkable ways. The contributions of Gibbs, Helmholtz, and van't Hoff are combined in the derivation shown in Box 5.2.

Box 5.1 Some foundational ideas for understanding the derivation in Box 5.2

Differentiation

There are a number of different ways of thinking about the expression, dG/dP. One can think of it as: the ratio of an infinitesimally small amount of free energy, dG, to an infinitesimally small amount of pressure, dP; the limit of the ratio, $\Delta G/\Delta P$, as ΔP approaches zero; the slope of the tangent to a graph of G against P at some nominated point; as a mathematical instruction to differentiate the function G with respect to the variable P, $d/dP(G)$; and as a measure of the rate at which G changes as P changes.

If $y = uv$ where u and v are functions of a variable 'x', then, $dy/dx = v\,du/dx + u\,dv/dx$, which is known as the product rule for differentiation. In thermodynamics one often writes this as, $dy = v\,du + u\,dv$. The reader will observe this in Box 5.2. The basic rule for differentiating a polynomial function, $y = ax^n$, where 'a' and 'n' are constants is $dy/dx = a.n.x^{n-1}$.

Integration

There are a number of different ways of thinking about the expression, $\int V\,dP$, the integral of V with respect to P. One can think of it as: the sum of the areas of rectangles constructed under the graph of V against P of height V and width dP as dP approaches zero; and as a mathematical operation which is the inverse of differentiation. If $y = ax^n$, where 'a' and 'n' are constants, then $\int y\,dx = \int ax^n dx = ax^{n+1}/(n+1) + c$ where c is a constant. The derivative of this answer should be ax^n.

An integral of relevance to Box 5.2 is:

$$\int_{P_1}^{P_2} \frac{1}{P}\,dP = [\ln P]_{P_1}^{P_2} = \ln P_2 - \ln P_1 = \ln \frac{P_2}{P_1}.$$

For redox reactions the standard free energy change, ΔG°, is related to the standard reduction potential, E^o, through $\Delta G^\circ = -nFE^\circ$, where E^o can be thought of as the work involved in driving 1 C of charge through the electrical circuit; F is the Faraday constant 96,487 C/mol; and n is the moles of electrons involved. Thus E° can be thought of as equivalent to Bergman's elective affinity according to Lindauer [22], remembering that the magnitude of oxidation potentials is the same as that of reduction potentials. The fact that $\Delta G°$ is also equal to $(-RT \ln K)$ suggests that the equilibrium constant, K, is also a measure of the 'elective affinity' of Bergman. Activity, as the product of the activity coefficient and the concentration, can be thought of as Berthollet's contribution to chemical affinity. So the concept of elective affinity was finally quantified in terms of the equilibrium constant and, in the case of redox reactions, by standard reduction potentials.

Box 5.2 The derivation of the driving force equation, the free energy equation, for a chemical reaction

The Gibbs function is defined as : $G = H - TS$

$$\text{Therefore :} \quad dG = dH - d(\text{TS})$$
$$dG = dH - SdT - TdS$$
$$dG = d(U + PV) - SdT - TdS$$
$$dG = dU + d(PV) - SdT - TdS$$
$$dG = dq + dw + VdP + PdV - SdT$$
$$- TdS(\Delta U = q + w; dU = dq + dw)$$
$$dG = TdS - PdV + VdP$$
$$+ PdV - SdT - TdS$$
$$dG = VdP - SdT$$

At constant temperature and using partial derivatives (often used when one variable is held constant) one can deduce:

$$\left(\frac{\partial G}{\partial P}\right)_T = V$$

And when pressure is held constant: $\left(\frac{\partial G}{\partial T}\right)_P = -S$
Considering the gas state and the first derivative: $\left(\frac{\partial G}{\partial P}\right)_T = V$
At constant temperature one can write: $dG = VdP$
Integrating from P_1 to P_2: $\int_{P_1}^{P_2} dG = \int_{P_1}^{P_2} \frac{nRT}{P} dP$, making use of the ideal gas law.

$$G(P_2) - G(P_1) = nRT \ln \frac{P_2}{P_1}$$

If P_1 is taken as the pressure under standard conditions, 1 bar pressure at 25 °C:

$$G(P) = G(P^0) + nRT \ln \frac{P}{P^0}$$

$$\frac{1}{n}G(P) = \frac{1}{n}G(P^0) + RT \ln a,$$

$$\mu = \mu^0 + RT \ln a,$$

where μ is the molar free energy or chemical potential, and 'a' is the activity. For solutions, $a = m/m^o$, where m is the molality (mol kg^{-1}) and m^o is 1 mol kg^{-1}.

Considering a general chemical reaction: $2A + 3B \rightarrow C + 2D$

$$\Delta G = \sum G(\text{products}) - \sum G(\text{reactants})$$
$$\Delta G = (\mu_C + 2\mu_D) - (2\mu_A + 3\mu_B)$$

$$\Delta G = \mu_C^0 + RT \ln a_C + 2\mu_D^0 + RT \ln a_D^2 - 2\mu_A^0$$
$$- RT \ln a_A^2 - 3\mu_B^0 - RT \ln a_B^3$$
$$\Delta G = \{(\mu_C^0 + 2\mu_D^0) - (2\mu_A^0 + 3\mu_B^0)\} + RT \ln a_C$$
$$+ RT \ln a_D^2 - RT \ln a_A^2 - RT \ln a_B^3$$

$$\Delta G = \Delta G^0 + RT \ln \frac{a_C . a_D^2}{a_A^2 . a_B^3}$$

$$\Delta G = \Delta G^0 + RT \ln Q$$

The foregoing can be illustrated by referring to the metal displacement reactions previously discussed and suggested by Geoffroy's table of affinities in 1718. For the following reaction inferred from the Table: $2AgNO_3(aq) + Pb(s) \rightarrow 2Ag(s) + Pb(NO_3)_2$, the redox potential for, $Ag^+ + e \rightarrow Ag$, is 0.80 V, and for $Pb^{2+} + 2e \rightarrow Pb$, is -0.13 V. One can say that Ag^+ has a higher potential or affinity for gaining electrons than does Pb^{2+} given the magnitudes of the voltages. These voltages still apply even if one multiplies an equation throughout by two for example. The standard potential for the reaction as given is 0.93 V, the positive value indicating that one would expect Pb to displace Ag from silver nitrate. For the second reaction discussed previously, $Cu(NO_3)_2(aq) + Fe(s) \rightarrow Cu(s) + Fe(NO_3)_2(aq)$, the redox potential for, $Cu^{2+} + 2e \rightarrow Cu$, is 0.34 V, and for $Fe^{2+} + 2e \rightarrow Fe$, is -0.44 V. Thus Cu^{2+} has a stronger affinity for electrons than does Fe^{2+} given the values of the standard reduction potentials. The potential for the total reaction, 0.78 V, indicates that iron will displace Cu from copper nitrate. On further comparison between Geoffroy's Table and a Table of standard reduction potentials, one would expect lead and copper to exchange places in the Table of Affinities. However, the difficulty of obtaining pure samples of metals in 1718 would have made the task more onerous.

The iron(III) thiocyanate reaction is not a redox reaction so one cannot use standard reduction potential tables to quantify affinity. But one can use standard free energy data and equilibrium constant data. The paper by Laurence [24] in Table 1.1 calculates $\Delta G°$ for the reaction, $Fe^{3+} + SCN^- \rightleftharpoons FeSCN^{2+}$, to be -17.2 kJ/mol, suggesting that under the standard conditions at 298 K, the reactants have more free energy than the products by 17.2 kJ/mol and so one would expect the forward reaction to be spontaneous under these conditions. This is comparable to the value of $\Delta G°$ obtained by Betts and Dainton [25] of -16.7 kJ/mol. Much of the research shown in Table 1.1 revolves around the determination of equilibrium constants as a quantification of affinity and this will be our major concern in Chap. 6.

References

1. Quílez J (2018) A historical/epistemological account of the foundation of the key ideas supporting chemical equilibrium theory. Found Chem. https://doi.org/10.1007/s10698-018-9320-0
2. Knight D (1992) Ideas in chemistry. The Athlone Press, London, p 114
3. Geoffroy EF (1718) Table des Differents Rapports Observés en Chimie entre Differentes Substances. Mem Acad R Sci 202–212
4. de Berg K (2014) Teaching chemistry for all its worth: the interaction between facts, ideas, and language in Lavoisier's and Priestley's chemistry practice: the case of the study of the composition of air. Sci Educ 23(10):2045–2068
5. Gingras Y (2001) What did mathematics do to physics? Hist Sci xxxix:383–416
6. Geoffroy EF (1718) Table des Differents Rapports Observés en Chimie entre Differentes Substances. Mem Acad R Sci, p 202
7. Waage P, Guldberg CM (1864) Forhandlinger: Videnskabs-Selskabet I Christiana (trans: Studies concerning affinity by Abrash HI). In: Bastiansen O (ed) The law of mass action, a centenary volume. Universitetsforlaget, Oslo, p 35
8. Gladstone JH (1855) On circumstances modifying the action of chemical affinity. Phil Trans R Soc Lond 145:179–223
9. Waage P, Guldberg CM (1864) Forhandlinger: Videnskabs-Selskabet I Christiana (trans: Studies concerning affinity by Abrash HI). In: Bastiansen O (ed) The law of mass action, a centenary volume. Universitetsforlaget, Oslo, p 4
10. Waage P, Guldberg CM (1864) Forhandlinger: Videnskabs-Selskabet I Christiana (trans: Studies concerning affinity by Abrash HI). In: Bastiansen O (ed) The law of mass action, a centenary volume. Universitetsforlaget, Oslo, p 5
11. de Berg K, Maeder M, Clifford S (2017) The thermodynamic formation constants for iron(III) thiocyanate complexes at zero ionic strength. Inorg Chim Acta 446:249–253
12. Giere RN (1988) Explaining science: a cognitive approach. The University of Chicago Press, Chicago
13. Giere RN (1999) Science without laws. University of Chicago Press, Chicago
14. Lederman NG, Abd-El-Khalick F, Bell RL, Schwartz RS (2002) Views of nature of science questionnaire: toward valid and meaningful assessment of learners' conceptions of nature of science. J Res Sci Teach 39(6):497–521
15. Horner JK, Rubba PA (1979) The laws are mature theories fable. Sci Teach 46(2):31
16. Ghirardi M, Marchetti F, Pettinari C, Regis A, Roletto E (2014) A teaching learning sequence for learning the concept of chemical equilibrium in secondary school education. J Chem Educ 91(1):59–65
17. Driscoll DR (1979) Invitation to enquiry: the Fe^{3+}/CNS^- equilibrium. J Chem Educ 56(9):603
18. Lund EW (1965) Guldberg and Waage and the law of mass action. J Chem Educ 42(10):548–550
19. Lewis GN, Randall M (1923) Thermodynamics and the free energy of chemical substances. McGraw Hill, New York

20. Lindauer MW (1962) The evolution of the concept of chemical equilibrium from 1775 to 1923. J Chem Educ 39(8):389
21. Lindauer MW (1962) The evolution of the concept of chemical equilibrium from 1775 to 1923. J Chem Educ 39(8):390
22. Lindauer MW (1962) The evolution of the concept of chemical equilibrium from 1775 to 1923. J Chem Educ 39(8):384–390
23. Laidler KJ (1993) The world of physical chemistry. Oxford University Press, Oxford, p 84
24. Laurence GS (1956) A potentiometric study of the ferric thiocyanate complexes. Trans Faraday Soc 52:236–242
25. Betts RH, Dainton FS (1953) Electron transfer and other processes involved in the spontaneous bleaching of acidified aqueous solutions of ferric thiocyanate. J Am Chem Soc 75:5721–5727

Chapter 6
The Reaction and Its Equilibrium Constants: The Role of Mathematics and Data Analysis

There have been attempts at the secondary and tertiary levels of education to teach chemistry and physics without the use of mathematical equations and their derivations, supposedly simplifying the subject and making it more acceptable to students. But if our concern is with the *nature of chemistry*, one must realize that many of its laws are mathematical in nature. In fact, chemists had this outcome in mind since the 18th century as mathematics could add an exactness to chemistry which was missing in earlier days.

6.1 The Frank and Oswalt Technique

Eight of the papers listed in Table 1.1 used the mathematical technique of Frank and Oswalt [1] to determine the equilibrium constant K for the formation of the 1:1 complex, so it is worthwhile considering what this approach involved. Assuming that only a 1:1 complex forms from initial concentrations 'a' and 'b' from which a concentration 'x' reacts:

$$Fe^{3+}(aq) + SCN^-(aq) \rightleftharpoons FeSCN^{2+}(aq)$$

Initial conc.	a	b	0
React. conc.	$-x$	$-x$	$+x$
Equil. conc.	$(a-x)$	$(b-x)$	x

$$K(\text{using concentrations}) = \frac{x}{(a-x)(b-x)}$$

From spectrophotometric absorbance data at 460 nm and assuming that only $FeSCN^{2+}$ contributes to the absorbance and the Beer-Lambert law (absorbance of light passing through a solution is proportional to the concentration of the absorbing

© The Author(s), under exclusive license to Springer Nature Switzerland AG 2019
K. C. de Berg, *The Iron(III) Thiocyanate Reaction*, SpringerBriefs in History of Chemistry, https://doi.org/10.1007/978-3-030-27316-3_6

species in the solution) is satisfied for a 1 cm cell, that is, absorbance A = molar absorptivity times concentration = εx:

$$K = \frac{\frac{A}{\varepsilon}}{\left(a - \frac{A}{\varepsilon}\right)\left(b - \frac{A}{\varepsilon}\right)}$$

$$K\left[ab - (a+b)\frac{A}{\varepsilon} + \left(\frac{A}{\varepsilon}\right)^2\right] = \frac{A}{\varepsilon}$$

$$ab = \frac{A}{\varepsilon K} + (a+b)\frac{A}{\varepsilon} - \left(\frac{A}{\varepsilon}\right)^2$$

$$\frac{ab}{A} = \frac{(a+b)}{\varepsilon} + \frac{1}{\varepsilon K}$$

which is the equation of a straight line where $(A/\varepsilon)^2$ is considered to be so small as to make effectively zero contribution. If the assumptions are correct, then a plot of (ab/A) against $(a+b)$ should yield a straight line of slope $(1/\varepsilon)$ and intercept $(1/\varepsilon K)$. As an illustration of this, if one takes the data from Lister and Rivington [2], the following plot (Fig. 6.1) is obtained for data at 25 °C, 1 cm path length, and ionic strength $\mu = 1.2$ M. Ionic strength, μ, is a measure of the solution's electrostatic interaction capability and is defined as,

$$\mu = \left(\frac{1}{2}\right)\sum_i c_i z_i^2,$$

where c_i is the concentration of species, i, and z_i is its charge number. The slope leads to a molar absorptivity ε value of 4694 L mol^{-1} cm^{-1} and the intercept leads to a K value of 124.5 M^{-1}. The 'a' values varied from 0.01 to 0.1667 M for a 'b' value of 0.000125 M. Under these conditions one would only expect the 1:1 complex to

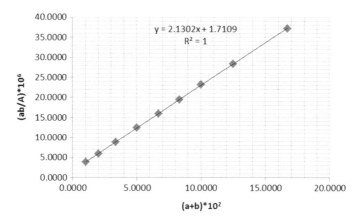

Fig. 6.1 The Frank-Oswalt plot using the spectrophotometric results of Lister and Rivington [2, 3]

form and the linear result obtained suggests the assumptions were valid. One must remember, of course, that chemists in the 1940s and 1950s would not have had access to laptop computers and Excel spreadsheets and would most likely have had to construct the graphs by hand on graph paper whereas Fig. 6.1 was constructed on an Excel spreadsheet. The fact that the regression equation is given by Excel also simplifies the procedure for finding K and ε.

The argument here is that the derivation of the equation,

$$\frac{ab}{A} = \frac{(a+b)}{\varepsilon} + \frac{1}{\varepsilon K},$$

is intimately connected to the chemistry and to focus only on the use of the final equation is to deny a familiarity with the chemical assumptions made in the derivation. These assumptions include: SCN^- is the only competitor for ferric ions; $FeSCN^{2+}$ is the only complex responsible for the red colour; Fe^{3+} and SCN^- make no contribution to the absorbance; the absorbance of the solution obeys the Beer-Lambert law; and that the molar absorptivity, ε, is large enough so that,

$$\left(\frac{A}{\varepsilon}\right)^2 \ll \frac{A}{\varepsilon K} + (a+b)\frac{A}{\varepsilon}.$$

6.2 Approximations and Iterations

When more than one species contributes to the absorbance, the complexity of the system grows in such a way that chemists have had to rely on approximations to achieve a solution for the equilibrium constants. Lister and Rivington [2] derive an equation as follows for the case where $FeSCN^{2+}$ and $Fe(SCN)_2^+$ contribute to the absorbance.

$$Fe^{3+} + SCN^- \rightleftharpoons FeSCN^{2+} \quad K_1 = \frac{[FeSCN^{2+}]}{[Fe^{3+}][SCN^-]}$$

$$FeSCN^{2+} + SCN^- \rightleftharpoons Fe(SCN)_2^+ \quad K_2 = \frac{[Fe(SCN)_2^+]}{[FeSCN^{2+}][SCN^-]}$$

$$\begin{aligned}
a = \text{total iron} &= [Fe^{3+}] + [FeSCN^{2+}] + [Fe(SCN)_2^+] \\
&= [Fe^{3+}] + K_1[Fe^{3+}][SCN^-] + K_2[FeSCN^{2+}][SCN^-] \\
&= [Fe^{3+}] + K_1[Fe^{3+}][SCN^-] + K_1 K_2[Fe^{3+}][SCN^-]^2 \\
&= [Fe^{3+}]\{1 + K_1[SCN^-] + K_1 K_2[SCN^-]^2\}
\end{aligned}$$

Let the absorbance at 460 nm $= A_{460} = \varepsilon_1[FeSCN^{2+}] + \varepsilon_2[FeSCN)_2{}^+]$

$$= \varepsilon_1 K_1[Fe^{3+}][SCN^-] + \varepsilon_2 K_1 K_2[Fe^{3+}][SCN^-]^2$$

$$= [Fe^{3+}]\{\varepsilon_1 K_1[SCN^-] + \varepsilon_2 K_1 K_2[SCN^-]^2\}$$

$$A_{460} = \frac{a}{\{1 + K_1[SCN^-] + K_1 K_2[SCN^-]^2\}} \cdot \left\{\varepsilon_1 K_1\left[SCN^-\right] + \varepsilon_2 K_1 K_2\left[SCN^-\right]^2\right\}$$

$$\frac{A_{460}}{a} \cdot \{1 + K_1\left[SCN^-\right] + K_1 K_2[SCN^-]^2\} - \varepsilon_1 K_1\left[SCN^-\right] = \varepsilon_2 K_1 K_2\left[SCN^-\right]^2$$

The idea was to repeatably choose values of K_2 until $\varepsilon_2 K_1 K_2$ became constant. K_1 and ε_1 had been previously determined (Fig. 6.1) but how would $[SCN^-]$ be determined? If b could represent total thiocyanate, then:

$$[SCN^-] = b - \{[FeSCN^{2+}] + 2[FeSCN)_2{}^+]\}$$

As a first approximation Lister and Rivington [2, 3] took ε_2 to be $2\varepsilon_1$. Since

$$A_{460} = \varepsilon_1[FeSCN^{2+}] + \varepsilon_2[FeSCN)_2{}^+]$$

$$= \varepsilon_1[FeSCN^{2+}] + 2\varepsilon_1[FeSCN)_2{}^+]$$

$$= \varepsilon_1\{[FeSCN^{2+}] + 2[FeSCN)_2{}^+]\}$$

Therefore,

$$[SCN^-] = b - (A_{460}/\varepsilon_1)$$

Again, this must have proved a very tedious process before the advent of laptop computers and Excel spreadsheets. If one plots, $(A_{460}/a).\{1 + K_1[SCN^-] + K_1 K_2[SCN^-]^2\} - \varepsilon_1 K_1[SCN^-]$ against $[SCN^-]^2$, one can adjust the value of K_2 on a spreadsheet until the best straight line is obtained, the slope of which will correspond to $\varepsilon_2 K_1 K_2$ from which ε_2 can be obtained.

Lister and Rivington [3] contains data for total iron equal to 0.001 M for thiocyanate concentrations from 0.00025 to 0.25 M at 25 °C at an ionic strength of 1.2 M. The best straight line is shown in Fig. 6.2 for Lister and Rivington's data. The value of K_2 which gave the best straight line passing closest to the origin was 10.7 M^{-1} and the slope yielded an ε_2 value of 10,275 L mol^{-1} cm^{-1}.

Today, data analysts would not regard the manual adjustment of K_2 on a spreadsheet as the best way to find a value for K_2 and ε_2. A procedure making use of Excel's *Solver* would be regarded more favourably. The idea here is to compare the calculated absorbance from the expression,

$$A_{460} = \frac{a}{\{1 + K_1[SCN^-] + K_1 K_2[SCN^-]^2\}} \cdot \{\varepsilon_1 K_1[SCN^-] + \varepsilon_2 K_1 K_2[SCN^-]^2\},$$

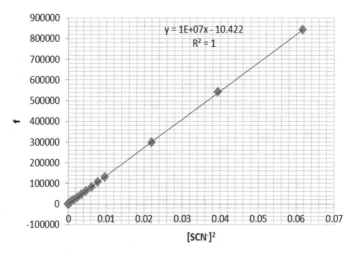

Fig. 6.2 The plot of 'f' against $[SCN^-]^2$ for the data from Lister and Rivington [2, 3] where 'f' refers to $\{\frac{A_{460}}{a}.\{1 + K_1[SCN^-] + K_1K_2[SCN^-]^2\} - \varepsilon_1 K_1[SCN^-]\}$

with the experimental absorbance, and then ask *Solver* to minimize the sum of the squares of the differences between the calculated and experimental absorbances by allowing values for K_2 and ε_2 to change in order to accomplish this. Applying *Solver* to Lister and Rivington's results leads to $K_2 = 12.32$ M^{-1} and $\varepsilon_2 = 9913$ L mol^{-1} cm^{-1}.

Again, the significance of the derivation of the equation for A_{460} lies in the chemical assumptions associated with the derivation, namely: no higher complexes than the 1:2 complex exist under the experimental conditions; the total absorbance at a particular wavelength can be expressed as the sum of the absorbances due to FeSCN^{2+} and Fe(SCN)$_2$$^+$; the Beer-Lambert law is obeyed by all coloured species at all concentrations; and the use of the equation, $[SCN^-] = b - (A_{460}/\varepsilon_1)$, only strictly applies if ε_2 is $2\varepsilon_1$. So there is an implicit error in how the chemist gains access to chemical constants like equilibrium constants and molar absorptivities from experimental data. The duty of the chemist is to try to reduce these errors to a minimum. One can see how critical mathematics and data analysis are in this process.

6.3 The Technique of Global Analysis

From about the 1980s sophisticated computer software analysis programs were beginning to be developed to cope with the simultaneous analysis of multiple equilibria with a much larger set of data than was possible previously [4]. The main distinction in the papers by de Berg et al. [5, 6] in Table 1.1 compared with the earlier methods used to obtain the data for determining K_1 and K_2 is the use of a complete spectral series of absorbances acquired at many wavelengths and analysed globally, rather than the use of analyses at individual wavelengths. In addition, no approximations are involved in the computations of the equilibria. The data analysis reported in these papers makes use of a more recent generalised and user-friendly adaptation called *ReactLab Equilibria* [7]. Compared to previous techniques, global analysis is more robust and results in estimates for the molar absorptivity spectra of all complexes as well as the equilibrium constants.

In the papers by de Berg et al. [5, 6] over 77,000 absorbance measurements (256 spectra at 301 wavelengths) were taken by the stopped-flow instrument for each admission of an iron(III) and thiocyanate solution. In order to determine an initial spectrum at time zero across the entire wavelength range to minimize the impact of the instability of the product, a pragmatic mechanism involving a very fast pseudo first-order reaction between iron(III) and thiocyanate ions to form an intermediate followed by a first-order decay to some unknown product P was applied [5]. The initial spectra across a range of thiocyanate concentrations for an initial ionic strength of 1.0 M are shown in Fig. 6.3. One can see that the absorbance maximum shifts from about 460 nm at low SCN^- concentrations to about 480 nm at higher SCN^- concentrations. The blood-red colour is retained over this wavelength range.

Having determined the initial spectra for all thiocyanate concentrations relevant to a particular ionic strength, equilibrium analyses were performed with *ReactLab Equilibria* to determine equilibrium constants and molar absorptivity data. The model used for the quantitative analysis of the equilibria was:

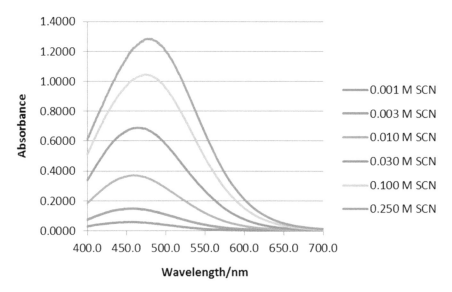

Fig. 6.3 Extrapolated spectra at zero time (initial spectra) for $[SCN^-]$ from 0.001 M (bottom) to 0.250 M (top), $[Fe^{3+}] = 1.5 \times 10^{-4}$ M at an initial ionic strength of 1.0 M

$$Fe^{3+} + SCN^- \xrightleftharpoons{K_1^\mu} FeSCN^{2+}$$

$$FeSCN^{2+} + SCN^- \xrightleftharpoons{K_2^\mu} Fe(SCN)_2{}^+$$

$$H^+ + SCN^- \xrightleftharpoons{K=7.94\times10^{-2}/M^{-1}} HSCN$$

$$Fe^{3+} + OH^- \xrightleftharpoons{3.548\times10^{11}/M^{-1}} FeOH^{2+}$$

The model includes the formation of the initial 1:1 complex and the secondary 1:2 complex. Competing reactions for thiocyanate ions and iron(III) include the acidification of thiocyanate ions and the hydrolysis of iron(III). K values for the acidification and hydrolysis reactions were obtained from the literature [8]. Researchers have attempted to determine K values under constant ionic strength conditions since we know ionic strength can affect the equilibrium concentrations. The addition of a non-reacting sodium perchlorate solution is often used to approximate constant ionic strength conditions from one thiocyanate concentration to the next. Values of K_1^μ and K_2^μ for an ionic strength $\mu = 0.5$ M are shown in Table 6.1 for a temperature of 25 °C. The data by de Berg et al. [5] makes use of an initial absorbance spectrum at time zero to minimize the impact of the instability of the complex shown by the fading of its colour. The K values are significantly less than those reported earlier where initial spectra were not used. The data in Table 6.1 are typical of many coordination complexes where the equilibrium constant for adding the second ligand is less than that for adding the first ligand. The reason given for this phenomenon is a statistical one, the probability of a second ligand colliding with an attachment point on the Fe^{3+} being lower than the probability of the first ligand finding an attachment point on Fe^{3+}.

Table 6.1 Values of K_1^μ and K_2^μ for the iron(III) thiocyanate reaction at 25 °C and ionic strength, μ, of 0.5 M

Year	Technique	K_1^μ/M^{-1}	K_2^μ/M^{-1}	Reference
1947	Spectrophotometry	138	–	[1][a]
1956	Potentiometric	139	20.4	[9]
1958	Spectrophotometry	145	14	[10][b]
1958	Potentiometric	133	10	[10][c]
1963	Spectrophotometry	169	–	[11][d]
1998	Spectrophotometry	146	–	[12]
2016	Spectrophotometry	98	7	[5]

[a]Temperature not specified
[b]Temperature was 18 °C and μ was 0.56
[c]Temperature was 20 °C and μ was 0.65
[d]Temperature was not specified

A consideration of log K values across a number of different ionic strengths can lead to an estimate of log K^0, a log K value at zero ionic strength. To do this one needs to use activities in place of concentrations where activity, $a = \gamma c$, where γ is the activity coefficient usually with values ≤ 1 and c is the concentration. Activity is related to the ionic strength usually in the form of an adaptation of the Debye-Hückel equation. The following relationship was used in the study by de Berg et al. [6]:

$$\log \gamma_i = -A z_i^2 \frac{\sqrt{\mu}}{\left(1 + \alpha \sqrt{\mu}\right)} + \beta \mu,$$

where α and β are fitted or fixed parameters and A has the value of 0.509 $(\mathrm{dm}^3/\mathrm{mol})^{1/2}$ for aqueous solutions. Substituting the activity formula into the appropriate K expressions leads to the following equations used to determine log K_1^0 and log K_2^0:

$$\log K_1^\mu = \log K_1^o - \frac{6\sqrt{\mu}}{\left(1 + 1.5\sqrt{\mu}\right)} - \beta_1 \text{ and } \log K_2^\mu = \log K_2^o - \frac{4\sqrt{\mu}}{\left(1 + 1.5\sqrt{\mu}\right)} - \beta_2.$$

The appropriate plots to find log K_1^0 and log K_2^0 are shown in Fig. 6.4. A summary of the log K_1^0 and log K_2^0 zero ionic strength equilibrium constant values obtained from 1947 to 2017 is shown in Table 6.2. One can see that the determination of log K_2^0 presented a major challenge to chemists and this can be seen in the mathematical challenges previously discussed.

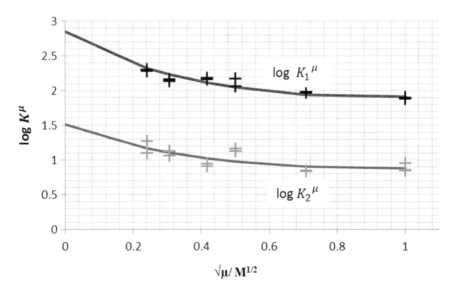

Fig. 6.4 The plots of experimental log K^μ points and calculated log K^μ curves against the square root of the average ionic strength $\sqrt{\mu}$ showing fitted intercepts for log K_1^0 of 2.85 and log K_2^0 of 1.51

Table 6.2 Published values of $\log K_1^0$ and $\log K_2^0$ based on experimental data and SI chemical data

Year	Temperature	$\log K_1^0$	$\log K_2^0$	Reference
1947	–	2.95	–	[1]
1953	25 °C	2.94	–	[13]
1955	25 °C	3.03	–	[2]
1956	25 °C	3.03	–	[9]
1958	18 °C	3.04	1.60	[10]
1998	–	2.96	–	[12]
2014	25 °C	3.02	–	[8]
2017	25 °C	2.85 ± 0.08	1.51 ± 0.13	[6]

It is important to realize that the model doesn't naturally emerge from the absorbance data but, from one's previous chemical knowledge, needs to be specified before any global analysis of the data commences. So, in a sense, the calculated equilibrium and molar absorptivity data are only as good as the input model. If the calculated absorbance data is a good fit to the experimental absorbance data, one's confidence in the input model and its equilibrium data grows. If a good fit does not emerge from the computer calculations then one might call into question the input model. One can see in Fig. 6.5 that the calculated absorbance points match the experimental absorbance curve very well indicating support for the equilibrium model proposed. The global analysis program also depends on initial estimates of the K and ε values being made. There is a sense in which one has to assume what

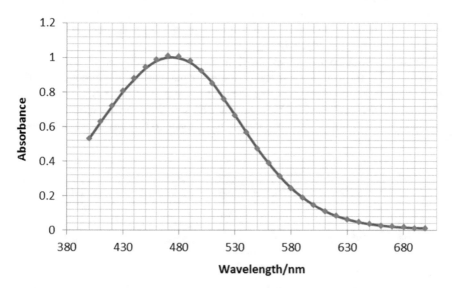

Fig. 6.5 The plot of absorbance against wavelength for $[Fe^{3+}]$ of 1.5×10^{-4} M and $[SCN^-]$ of 0.1 M showing calculated absorbance points in relation to the experimental absorbance curve

the answer should look like before one does a detailed calculation of it. This might seem a strange way of accumulating chemical knowledge but it must always position itself in reference to experimental data as shown in Fig. 6.5. Hoffmann [14] calls this kind of approach "nearly circular reasoning." While a case has been made for familiarising oneself with the mathematics associated with calculating values for K and ε, one cannot be so expectant when it comes to accessing the mathematics behind complicated software programs. However, one should be able to articulate what a software program is accomplishing in relation to the data and this will be easier to do if one has had experience with the mathematics of simplified systems discussed earlier.

6.4   Instrumentation

The role of instrumentation has been critical in producing values for the equilibrium constants. The focus here will be on the visible spectrophotometer, the major instrument used in the work reported in the papers of Table 1.1. Colour intensities before 1940 were largely compared by looking vertically down two tubes placed in the path of diffuse light and adjusting the height of the experimental solution until the colour matched that of a standard solution. Gladstone [15] recognized how large the errors can be in such a process and always depended on his research assistant to analyse the colour intensities to reduce the bias he would automatically bring to the study. The invention of the modern spectrophotometer is usually attributed to Arnold Beckman in 1940 while working at his National Technologies Laboratory. Over the next 40 years improvements such as the use of a quartz prism rather than a standard glass prism and in wavelength resolution were made. The basic components of the spectrophotometer are shown in Fig. 6.6.

Microprocessor-controlled spectrophotometers were produced by Cecil Instruments in 1981. The resulting automation improved the speed of measurement. Double beam versions of the instrument were developed from 1984 to 1985. Connection to external software that provided PC control and on-screen displays of spectra occurred in the 1990s, and improvements continue to be made to this day. Developments in

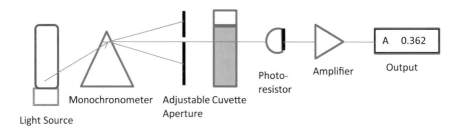

Fig. 6.6 Basic components of a single beam spectrophotometer

Fig. 6.7 Small volumes of solution are driven from syringes A and B into the mixer from which the mixture enters the absorbance cell at which time syringe C is driven to stop the flow before absorbance measurement

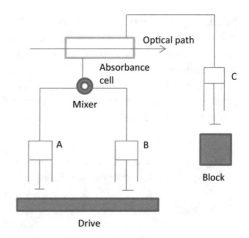

optics, electronics, and computing were responsible for the improvements. Today, it is one of the most common analytical instruments used in analytical laboratories.

The combination of the spectrophotometer with a stopped-flow mechanism enabled chemists to measure the rate of very fast reactions like the iron(III) thiocyanate reaction and to establish an initial spectrum for iron(III) thiocyanate species before fading of the colour commences. To the naked eye the reaction appears to be instantaneous but the stopped-flow mechanism can measure a spectrum within the millisecond range. Six of the studies listed in Table 1.1 used this technique. The stopped-flow mechanism is illustrated in Fig. 6.7. The volume injected, which is usually quite small, is limited by the stop syringe C which provides the "stopped-flow". Just prior to stopping, a steady state flow is achieved. The solution entering the absorbance cell is only milliseconds old. The age of this reaction volume is also known as the dead time of the stopped-flow system. As the solution fills the stopping syringe, the plunger hits a block, causing the flow to be stopped instantaneously. Using appropriate techniques, the kinetics of the reaction can be measured in the cell. Reaction times of the order of milliseconds can be measured. Attempts at measuring the kinetics of the iron(III) thiocyanate reaction will form the content of Chap. 7.

6.5 The Potentiometric Method

The other method used by some of the studies shown in Table 1.1 for determining equilibrium constants was the Potentiometric Method. The method used by Laurence [9] will serve as an example. The apparatus used was an electrochemical cell shown in Box 6.1 consisting of a standard cell on the left-hand side and a titration cell on the right-hand side where the iron(III) thiocyanate reaction takes place as aqueous ammonium thiocyanate is progressively added from a burette. Gold electrodes were used in both half-cells. The equilibrium between iron(III) and iron(II) in both half cells is responsible for the voltage of the cell for the cell reaction: $Fe^{3+} \rightleftarrows Fe_0^{3+}$. Fe_0^{3+} is the free ferric ion in the absence of thiocyanate ions and Fe^{3+} is the free ferric ion in the presence of thiocyanate ions. The form of the Nernst Equation used for determining Fe^{3+} was: $\ln\{[Fe^{3+}]/[Fe_0^{3+}]\} = -EF/RT$, where E is the voltage of the cell, R is $8.314\,J\,K^{-1}\,mol^{-1}$, F is Faraday's constant 96,487 C/mol, and T is the kelvin temperature. It is worth remembering that this equation is just the electrochemical equivalent of the driving force equation introduced in the previous chapter,

---

Box 6.1 Cell notation for a potentiometric cell consisting of a standard half-cell on the left of a salt bridge and a titration half-cell on the right into which NH$_4$SCN is added from a burette and the iron(III) thiocyanate complexes form

$$Au \ \left|Fe_0^{3+}(aq)/Fe^{2+}(aq)\right|\left|Fe^{3+}(SCN)/Fe^{2+}(SCN^-)\right|\ Au$$

$$\Delta G = \Delta G^o + RT \ln \frac{[Fe_0^{3+}]}{[Fe^{3+}]},$$

where $\Delta G = -nFE$ and $\Delta G^o = -nFE^o$, and E is the work done in joules in moving one coulomb of charge through the circuit which just represents the voltage.

The equilibrium model used in the data analysis consisted of the following three equilibria:

$$Fe^{3+} + SCN^- \rightleftharpoons FeSCN^{2+} \quad K_1 \tag{6.1}$$

$$FeSCN^{2+} + SCN^- \rightleftharpoons Fe(SCN)_2^+ \quad K_2 \tag{6.2}$$

$$Fe^{3+} + H_2O \rightleftharpoons FeOH^{2+} + H^+ \quad K_H \tag{6.3}$$

In the left-hand cell:

$$[Fe_0^{3+}] = [Fe_t^{3+}]/(1 + K_H/[H^+]) \tag{6.4}$$

where $[Fe_t^{3+}]$ is the total ferric ion concentration coordinated and free.
In the right-hand cell:

$$[Fe_c^{3+}] = [Fe_t^{3+}](1 - [Fe^{3+}]/[Fe_0^{3+}]) \tag{6.5}$$

where $[Fe_c^{3+}]$ is the ferric ion concentration coordinated to thiocyanate. Successive approximations and iterations were made using three equations derived from

Eqs. (6.1) to (6.5) as follows:

$$K_1 = [Fe_c^{3+}]/\{[Fe^{3+}]([SCN_t^-] - [Fe_c^{3+}])\} \tag{6.6}$$

$$[SCN^-] = [SCN_t^-] - [Fe_c^{3+}](1 + 2K_2[SCN^-])/(1 + K_2[SCN^-]) \tag{6.7}$$

$$K_2 = [Fe_c^{3+}]/([Fe^{3+}][SCN^-]^2 K_1) - 1/[SCN^-] \tag{6.8}$$

Here, $[SCN^-]$ is the free concentration of thiocyanate approximated to $\{[SCN_t^-] - [Fe_c^{3+}]\}$. Equation (6.6) neglects Eq. (6.2) as a first approximation.

The iteration and approximation procedure began by plotting the K_1 values determined from Eq. (6.6) against $[SCN^-]$ for each voltage of the cell and extrapolating to zero thiocyanate concentration. A plot for Laurence's data at 25 °C, ionic strength 0.5 M, $[Fe_t^{3+}]$ of 8.07×10^{-4} M and $[H^+]$ of 4.1×10^{-2} M is shown in Fig. 6.8. From the plot an initial value of K_1 of 159 M^{-1} was chosen for the iteration. Using this K_1 value and Eq. (6.8) a value for K_2 was calculated for each of the ten experimental points. Using the K_2 values calculated, Eq. (6.7) was used to find corrected values for the free thiocyanate concentration, $[SCN^-]$. These corrected values for $[SCN^-]$ were substituted into Eq. (6.8) to find better estimates for K_2 and then Eq. (6.7) was used to find better values for $[SCN^-]$. This iteration/approximation procedure continued until $[SCN^-]$ converged so no change was noted. This happened within three iterations. The value of K_1 was changed until the range of K_2 values reached a minimum. Using the facility of Excel and a K_H value of 6.76×10^{-3}, the value of K_1 was 153 M^{-1} and K_2 was 23.2 M^{-1}. In 1956, an accurate value for K_H was not known and Laurence used a value obtained by Bray and Hershey in 1934, which was probably 1.8×10^{-3} for ionic strengths between 0.36 and 1 [16]. For the concentrations and voltages used for Fig. 6.8, Laurence determined K_1 to be 139 M^{-1} and

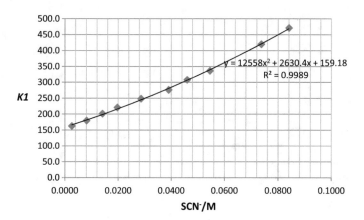

Fig. 6.8 The plot of calculated K_1 values against the concentration of free thiocyanate ions approximated using $\{[SCN_t^-] - [Fe_c^{3+}]\}$

K_2 to be 20.9 M^{-1}. Apart from the fact that Laurence would not have had EXCEL available to plot and minimize differences and did not have an accurate value for K_H, a titration lasting nearly an hour and a half must have been affected by the fading of the blood-red colour and consequently the [Fe^{3+}] if indeed the fading is due to reduction of Fe^{3+} to Fe^{2+}. In fact this will be a drawback for all titration techniques for studying this reaction. It has been found that, for the spectrophotometric technique, the longer one waits to take absorbance measurements the more inflated the calculated K values [5]. This is borne out in Tables 6.1 and 6.2 of this chapter.

Laurence also used the van't Hoff law,

$$\ln K = \frac{-\Delta H}{RT} + \text{constant},$$

to determine the enthalpy change, ΔH, for reactions (6.1) and (6.2) from the slope of the plot of, $\ln K$ against 1/T, as shown in Fig. 6.9. Multiplying the slope by R (8.314 J K^{-1} mol^{-1}) gives ΔH for reaction (6.1) of -6.2 kJ mol^{-1} and for reaction (6.2) of -1.2 kJ mol^{-1}. The van't Hoff law illustrates again the power that differential and integral calculus was to bring to chemistry. The law is derived from,

$$\left(\frac{\partial G}{\partial T}\right)_P = -S,$$

which was derived from, $G = H - TS$, in the previous chapter (Fig. 5.3). The van't Hoff derivation is given in Box 6.2 to illustrate again the important role of mathematics.

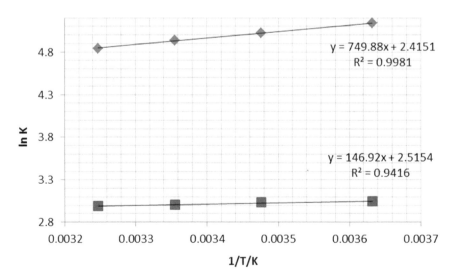

Fig. 6.9 The plot of ln K against 1/T for the reaction forming $FeSCN^{2+}$ (top line) and $Fe(SCN)_2{}^+$ (bottom line) using the data from Laurence [9]

In 2001 Gingras [17] asked the question, "What did mathematics do to physics?" I think one should also ask the question, "What has mathematics and data analysis done to chemistry?" Before mathematics changed the face of physics in the 18th century, physics belonged to the public domain and anyone who had an interest in making water wheels, lenses, levers and so on felt able to participate in experiments and debates about physical phenomenon. Once calculus invaded the subject of physics, it moved from the public domain to the private domain of professionals who possessed the skills to interpret what people like Newton were proposing. To some extent this has also happened in chemistry but about 200 years later. I don't think the impact has been as great in chemistry due probably to the diversity of fields in chemistry. Specialisation in physical chemistry demands significant mathematical skill but areas in organic and inorganic chemistry are still available to the non- mathematician and if some mathematical prowess is required it often comes in the form of software packages for computers which will perform the necessary calculations. As far as the iron(III) thiocyanate reaction is concerned, we have seen how the handling of multiple equilibria required some mathematical skill in approximations and iterations in the 1950s, but computer software packages are now available to do this task. But while advanced calculus may not be required of the medicinal chemist, organic chemist, or inorganic chemist, some basic mathematical skill is required across all fields. It is my argument that an appreciation for the rigour that mathematics has brought to chemistry is essential for understanding the nature of modern chemistry.

We have seen how informed guesswork or theorising has partnered with experimentation in the chemical models presented for the iron(III) thiocyanate behaviour in aqueous solution. This has been seen in the spectrophotometric and potentiometric techniques. Knight [18] says of chemistry, "it has always been a central kind of science, useful in all sorts of ways, and from the beginning it was based upon experiment informed by theorizing in a way other sciences were not." This has often led to controversy as described earlier between Bergman and Berthollet who interpreted the same experimental evidence differently. But over time this controversy led to progress in that both of these great chemists eventually made important contributions to chemistry. As far as chemical modelling is concerned, does the inclusion of the formation of a higher complex like $[Fe(SCN)_2{}^+]$ improve the match between predicted absorbances and experimental absorbances. The fact that it does in spectrophotometry lends support to its existence as well as a shift in the absorbance maximum from 460 to 480 nm. The delicate blend between theory and experiment is evident here.

Box 6.2 Derivation of the van't Hoff law showing that a plot of ln K against $1/T$ should yield a straight line of slope $-(\Delta H^o/R)$ provided ΔH^o is constant over the temperature range

Consider

$$\left(\frac{\partial G}{\partial T}\right)_P = -S$$

$$dG = -S dT = \frac{G - H}{T} dT \quad \text{(from } G = H - TS\text{)}$$

Now:

$$\frac{d}{dT}\left(\frac{G}{T}\right) = \frac{1}{T}\frac{dG}{dT} + G d\left(\frac{1}{T}\right) dT \quad \text{(product rule)}$$

$$= \frac{1}{T}\frac{dG}{dT} + G\left(\frac{-1}{T^2}\right)$$

$$= \frac{1}{T}\frac{dG}{dT} - \frac{G}{T^2}$$

$$= \frac{G - H}{T^2} - \frac{G}{T^2} \quad \text{(using the expression on the second line)}$$

$$= \frac{-H}{T^2}$$

Therefore:

$$\frac{d}{dT}\left(\frac{\Delta G^o}{T}\right) = \frac{-\Delta H^o}{T^2} \quad \text{(applying to a chemical reaction under standard conditions)}$$

$$\frac{d}{dT}(-R\ln K) = \frac{-\Delta H^o}{T^2}$$

$$\frac{d\ln K}{dT} = \frac{\Delta H^o}{RT^2}$$

$$d\ln K = \frac{\Delta H^o}{RT^2} dT$$

Therefore:

$$\int d\ln K = \int \frac{\Delta H^o}{RT^2} dT \quad \text{(considering the indefinite integral)}$$

$$\ln K = \frac{-\Delta H^o}{RT} + \text{constant} \quad \text{(one form of the van't Hoff law)}$$

References

1. Frank HS, Oswalt RL (1947) The stability and light absorption of the complex ion $FeSCN^{++}$. J Am Chem Soc 69:1321–1325
2. Lister MW, Rivington DE (1955) Some measurements on the iron(III)-thiocyanate system in aqueous solution. Can J Chem 33(10):1572–1590
3. Lister MW, Rivington DE (1955) Some measurements on the iron(III)-thiocyanate system in aqueous solution. Can J Chem 33(10):1579
4. Maeder M, Neuhold YM (2007) Practical data analysis in chemistry. Elsevier, Amsterdam
5. de Berg K, Maeder M, Clifford S (2016) A new approach to the equilibrium study of iron(III) thiocyanates which accounts for the kinetic instability of the complexes particularly observable under high thiocyanate concentrations. Inorg Chim Acta 445:155–159
6. de Berg K, Maeder M, Clifford S (2017) The thermodynamic formation constants for iron(III) thiocyanate complexes at zero ionic strength. Inorg Chim Acta 446:249–253
7. http://www.jplusconsulting.com/
8. Blackman A, Gahan L (2014) SI chemical data, 7th edn. Wiley, Milton
9. Laurence GS (1956) A potentiometric study of the ferric thiocyanate complexes. Trans Faraday Soc 52:236–242
10. Perrin DD (1958) The ion $Fe(CNS)_2^+$. Its association constant and absorption spectrum. J Am Chem Soc 80(15):3852–3856
11. Ramette RW (1963) Formation of monothiocyanatoiron(III). J Chem Educ 40(2):71–72
12. Cobb CL, Love GA (1998) Iron(III) thiocyanate revisited. A physical chemistry equilibrium lab incorporating ionic strength effects. J Chem Educ 75(1):90–92
13. Betts RH, Dainton FS (1953) Electron transfer and other processes involved in the spontaneous bleaching of acidified aqueous solutions of ferric thiocyanate. J Am Chem Soc 75:5721–5727
14. Hoffmann R (2012) Nearly circular reasoning. In: Kovac J, Weisberg M (eds) Roald Hoffmann on the philosophy, art, and science of chemistry. Oxford University Press, New York, pp 45–52
15. Gladstone JH (1855) On circumstances modifying the action of chemical affinity. Phil Trans R Soc Lond 145:179–223
16. Bray WC, Hershey AV (1934) The hydrolysis of ferric ion. The standard potential of the ferric-ferrous electrode at 25 °C. The equilibrium $Fe^{+++} + Cl^- = FeCl^{++}$. J Am Chem Soc 56:1889–1893
17. Gingras Y (2001) What did mathematics do to physics? Hist Sci xxxix:383–416
18. Knight D (1992) Ideas in chemistry. The Athlone Press, London

Chapter 7
The Reaction and Its Kinetics

7.1 A Brief Historical Note

The rate of a chemical reaction first received serious study in 1850 when the German chemist Ludwig Wilhelmy (1812–1864) studied the rate of the inversion of sucrose using a polarimeter to follow the reaction at different concentrations of sucrose and acid [1, 2]. He measured the rate of change of the sugar concentration to be proportional to both the sugar and acid concentrations. This was the first time that a differential equation $(-dc/dt = kc)$ was integrated to obtain an expression for 'c' as a function of time, assuming the rate was proportional to concentration. The assumption was proved correct when the experimental results were consistent with the integrated equation, $\ln c = -kt + \text{constant}$. The rate of the reaction between ethanol and acetic acid to give ethyl acetate and water was studied by Bertholet and Saint-Gilles [3] and was found to be proportional to the concentration of both reactants. Harcourt and Esson [4, 5] studied the kinetics of the reaction between hydrogen peroxide and hydrogen iodide and also the reaction between potassium permanganate and oxalic acid by integrating a differential equation used to express the rate and checking the predicted variation of concentration with time against the experimental results. Van't Hoff [6] made a major contribution to the study of equilibrium and kinetics. Using integrated rate equations again, he showed that the decomposition of arsine into arsenic and hydrogen was a first order reaction, $(-dc/dt = kc)$, and the hydrolysis of ethyl acetate was a second order reaction, first order in each reactant, $(dc/dt = kc_A c_B)$. The term 'order of reaction' was introduced by Ostwald in 1887 [7].

Even though Waage and Guldberg [8] proposed that a reaction reaches equilibrium when the rates of the forward and reverse reactions are equal and used rate equations like the following for the forward reaction between two reactants:

$$v = \frac{dx}{dt} = k(p - x)^a (q - x)^b,$$

© The Author(s), under exclusive license to Springer Nature Switzerland AG 2019
K. C. de Berg, *The Iron(III) Thiocyanate Reaction*, SpringerBriefs in
History of Chemistry, https://doi.org/10.1007/978-3-030-27316-3_7

where 'x' is the amount reacted at time, 't', they were severely criticised for continuing to use the term 'chemical force' [9] to describe the forward and reverse reactions and Laidler [10] concludes that "they arrived at equilibrium equations on the basis of assumed kinetic equations, a procedure which is not satisfactory, and their investigations did not really contribute to the understanding of reaction rates." However, the work of Guldberg and Waage nevertheless "constituted an important milestone in the development of chemical equilibrium" and Ostwald and Nernst "praised their legacy as an epoch-making work" [11].

7.2   Reaction Rate Mechanism Proposed Since 1958

The reactions that were studied in the early years of chemical kinetics were slow enough to be timed with a good hand watch. However, the blood-red colour production of the iron(III) thiocyanate reaction is instantaneous to the naked eye and so a kinetic study did not appear until instruments were developed that could track fast reactions. A rapid mixing device was developed by Below et al. [12]; flash photolysis was used by Goodall et al. [13]; and stopped-flow was used by Gray and Workman [14] and Clark [15]. Stopped-flow was described in a previous chapter and a picture of the apparatus that was used by de Berg et al. [16, 17] is shown in Fig. 7.1. This equipment can measure reaction times of the order of milliseconds. A sample of 1.5×10^{-4} M ferric nitrate in 0.05 M HNO_3 and 0.003 M potassium thiocyanate were separately admitted to the stopped-flow apparatus shown in Fig. 7.1 and the absorbance of the mixture plotted against the time in milliseconds. The result is shown in Fig. 7.2. One can see that the iron(III) thiocyanate reaction has reached

Fig. 7.1 The stopped-flow apparatus showing the separate filling compartments for the reactants on the left and the stopping mechanism on the right

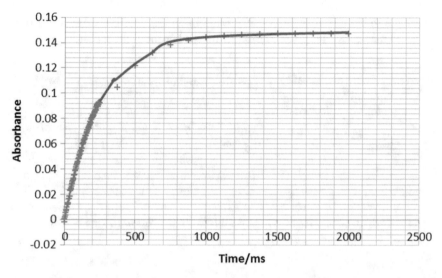

Fig. 7.2 The plot of Absorbance against Time in milliseconds for the reaction between 1.5×10^{-4} M Fe^{3+} in 0.05 M HNO_3 and 0.003 M SCN^- at 25 °C and ionic strength 1.0 M

equilibrium within one and a half seconds. Since the time of Below et al. experimental results have been interpreted in terms of a mechanism shown in Box 7.1 in the form expressed by Clark [15].

Box 7.1 A mechanism for the iron(III) thiocyanate reaction after Clark [15]

$$[Fe(OH_2)_6]^{3+} \quad + \quad SCN^- \quad \underset{k_{-1}}{\overset{k_1}{\rightleftharpoons}} \quad [Fe(OH_2)_5SCN]^{2+} \quad + \quad H_2O$$

$$\Big\Uparrow K_{a1} \qquad\qquad\qquad\qquad\qquad \Big\Uparrow K_{a2}$$

$$[Fe(OH_2)_5OH]^{2+} + \quad SCN^- \quad \underset{k_{-2}}{\overset{k_2}{\rightleftharpoons}} \quad [Fe(OH_2)_4(OH)SCN]^+ \quad + \quad H_2O$$

$$(+H^+) \qquad\qquad\qquad\qquad\qquad\qquad\qquad (+H^+)$$

The mechanism in Box 7.1 arose from a knowledge of the fact that a hydroxyl complex of Fe^{3+} was probably involved due to reaction rate increasing with pH. For the published kinetic studies since 1958 the iron concentration has exceeded the thiocyanate concentration by at least one to two orders of magnitude thus ensuring that $Fe(H_2O)_5SCN^{2+}$ is the species responsible for the blood-red colour, although Hoag [18] appears to allow for the possibility that $Fe(H_2O)_4(OH)SCN^+$ may also contribute to the colour. A brief background to writing rate law expressions from proposed mechanisms is given in Box 7.2 for the benefit of the reader.

Box 7.2 A brief background to writing rate law expressions given a mechanism

If a simple bimolecular collision between A and B to produce C is proposed: $A + B \rightarrow C$, then the rate law can be written as: Rate of disappearance of A, $-d[A]/dt = k[A][B]$, where k is the rate constant. In this case, $-d[A]/dt = -d[B]/dt = d[C]/dt$. If the mechanism proposed was: $2A \rightarrow B$, then $-d[A]/dt = k[A]^2 = 2d[B]/dt$. A first-order rate law can be expressed in differential form or in a form resulting from integration. In differential form it is: $-d[A]/dt = k[A]$. The term 'first order' refers to the power of the concentration which is one. If one expresses this in integrable form, then: $-d[A]/[A] = kdt$, and

$$\int_{[A]_0}^{[A]_t} \frac{-1}{[A]} d[A] = \int_0^t kdt.$$

The result of this integration is:

$$\{-\ln[A]\}_{[A]_0}^{[A]_t} = \{kt\}_0^t.$$

This leads to: $-\ln[A]_t - (-\ln[A]_0) = kt$. Simplifying gives:

$$\ln \frac{[A]_0}{[A]_t} = kt.$$

This is the form of the first-order rate law which will appear in subsequent derivations. A rate law expression can only be written from a proposed mechanism and not from the stoichiometric equation.

If a mechanism includes an equilibrium: $A + B \underset{k_{-1}}{\overset{k_1}{\rightleftharpoons}} C$, then $d[C]/dt = k_1[A][B] - k_{-1}[C]$.

The rate law expression for the mechanism given in Box 7.1 can be written in differential form as follows:

$$-\frac{d[Fe^{3+}]}{dt} = k_1[Fe(OH_2)_6^{3+}][SCN^-] + k_2[Fe(OH_2)_5OH]^{2+}[SCN^-]$$
$$- k_{-1}[Fe(OH_2)_5SCN]^{2+} - k_{-2}[Fe(OH_2)_4(OH)SCN]^+$$

The Fe^{3+} refers to iron not coordinated to thiocyanate. Removing the coordinated water and the charges, this law can be written more simply as:

$$-\frac{d[Fe]}{dt} = k_1[Fe][SCN] + k_2[FeOH][SCN] - k_{-1}[FeSCN] - k_{-2}[Fe(OH)SCN]$$

$$(7.1)$$

Using the equilibrium constants, K_{a1} and K_{a2}, shown for the mechanism in Box 7.1, Eq. (7.1) can be rearranged to give Eq. (7.2) as shown.

$$-\frac{d[Fe]}{dt} = \left(k_1 + \frac{k_2 K_{a1}}{[H]}\right)[Fe][SCN] - \left(k_{-1} + \frac{k_{-2} K_{a2}}{[H]}\right)[FeSCN] \qquad (7.2)$$

Equation (7.2) can be expressed in the form of Eq. (7.3).

$$-\frac{d[Fe]}{dt} = k_f[Fe][SCN] - k_r[FeSCN] \qquad (7.3)$$

where

$$k_f = k_1 + \frac{k_2 K_{a1}}{[H]} \quad \text{and} \quad k_r = k_{-1} + \frac{k_{-2} K_{a2}}{[H]}$$

If 'x' represents the displacement from the equilibrium position, then:

$$[SCN] = [SCN]_{eq} + x \quad \text{and} \quad [Fe] = [Fe]_{eq} + x \qquad (7.4)$$

and

$$[FeSCN] = [FeSCN]_{eq} - x \qquad (7.5)$$

At equilibrium, $-\frac{d[Fe]}{dt} = 0$, so $k_f[Fe]_{eq}[SCN]_{eq} = k_r[FeSCN]_{eq}$.

Substituting Eqs. (7.4) and (7.5) into Eq. (7.3) and allowing $[Fe]_{eq} \cong [Fe]_T$, the total iron concentration, leads to:

$$-\frac{d\{[Fe]_{eq} + x\}}{dt} = k_f[Fe]_T\{[SCN]_{eq} + x\} - k_r\{[FeSCN]_{eq} - x\}$$

$$= k_f[Fe]_T[SCN]_{eq} + k_f x[Fe]_T - k_r([FeSCN]_{eq} - x)$$

$$= k_r[FeSCN]_{eq} + k_f x[Fe]_T - k_r([FeSCN]_{eq} - x)$$

$$= (k_r + k_f[Fe]_T)x$$

$$-\frac{dx}{dt} = (k_r + k_f[Fe]_T)x \qquad (7.6)$$

that is, a pseudo first-order reaction if $[Fe]_T$ is close in value to $[Fe]_{eq}$ which is the case given the fact that iron exceeds the thiocyanate concentration in the published kinetic accounts since 1958. Substituting Eq. (7.5) into Eq. (7.6) leads to:

$$-\frac{d\{[FeSCN]_{eq} - [FeSCN]\}}{dt} = \left(k_r + k_f[Fe]_T\right)\{[FeSCN]_{eq} - [FeSCN]\}$$

Rearranging for integration will give:

$$\frac{d[FeSCN]}{\{[FeSCN]_{eq} - [FeSCN]\}} = \left(k_r + k_f[Fe]_T\right)dt \qquad (7.7)$$

Integrating both sides of Eq. (7.7) from $t = 0$ to $t = t$ leads to:

$$\ln\frac{[FeSCN]_{eq}}{\{[FeSCN]_{eq} - [FeSCN]\}} = \left(k_r + k_f[Fe]_T\right)t$$

Since absorbance, $(A = \varepsilon cl)$, where ε is molar absorptivity, c is concentration, and l is the light path length, is used to measure the $[FeSCN]$ and $[FeSCN]_{eq}$ then:

$$\ln\frac{A_{eq}}{\left(A_{eq} - A\right)} = \left(k_r + k_f[Fe]_T\right)t \qquad (7.8)$$

$$\ln\frac{A_{eq}}{\left(A_{eq} - A\right)} = k_{obs}t \text{ where } k_{obs} = \left(k_r + k_f[Fe]_T\right) \qquad (7.9)$$

If a plot of $\ln(A_{eq}/(A_{eq} - A))$ against t yields a straight line then the assumption that $[Fe]_T$ can be approximated to $[Fe]_{eq}$ holds and the slope of the line should yield the rate constant, $k_{obs} = (k_r + k_f[Fe]_T)$. It is difficult to illustrate this from the published kinetic papers since no absorbance-time data is given. However the data leading to the plot in Fig. 7.2 from our unpublished research can be used to make the appropriate plot remembering that thiocyanate concentration exceeds the iron concentration under these circumstances so that,

$$\ln\frac{A_{eq}}{\left(A_{eq} - A\right)} = k_{obs}t \text{ where } k_{obs} = \left(k_r + k_f[SCN]_T\right).$$

The plot is given in Fig. 7.3. The assumptions hold with a k_{obs} value of 4.0742 s^{-1}. I will henceforth use the data published by Mieling and Pardue [19] to illustrate the approach used to find the kinetic data for the iron(III) thiocyanate reaction since this publication provides adequate data for illustration purposes.

Mieling and Pardue used Fe^{3+} concentrations from 0.005 to 0.02 M for a set thiocyanate concentration of 2.0×10^{-4} M. Under these circumstances Eq. (7.8) can be used where the observed first-order rate constant, k_{obs}, is $\left(k_r + k_f[Fe]_T\right)$. A plot of k_{obs} against $[Fe]_T$ should yield a straight line of slope k_f and intercept k_r. The plot

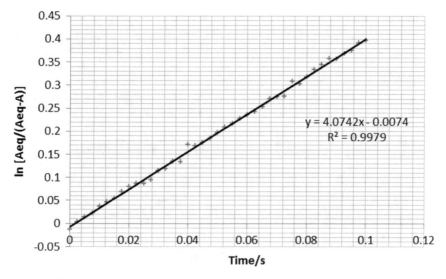

Fig. 7.3 The plot of $\ln\{A_{eq}/(A_{eq} - A)\}$ against time for initial concentrations of 1.5×10^{-4} M Fe^{3+}, 0.05 M H^+, and 0.003 M SCN^- and ionic strength of 1.0 M at 25 °C

for a set acid concentration of 0.225 M $HClO_4$ is shown in Fig. 7.4. The plot leads to a k_f value of 170 L mol^{-1} s^{-1} and to a k_r value of 1.34 s^{-1}. Repeating this plot for a range of acid concentrations will lead to a range of k_f and k_r values. A plot of k_f against $1/[H^+]$ should yield a straight line of slope, $k_2 K_{a1}$, and intercept k_1 according

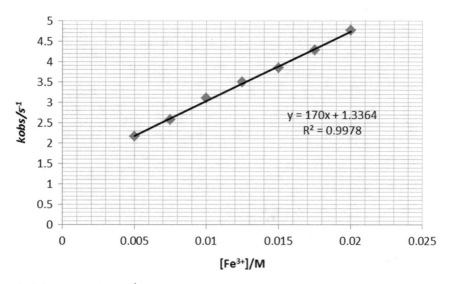

Fig. 7.4 The plot of k_{obs}/s^{-1} against the concentration of $Fe(ClO_4)_3/M$ for the iron(III) thiocyanate reaction at an acid concentration of 0.225 M $HClO_4$ after Mieling and Pardue [19]

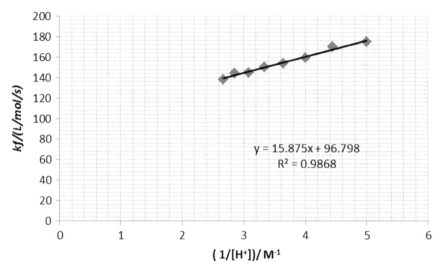

Fig. 7.5 The plot of k_f/L mol^{-1} s^{-1} against the reciprocal acid concentration for the iron(III) thiocyanate reaction at an iron concentration of 0.01 M after Mieling and Pardue [19]

to Eq. (7.3) and this is shown in Fig. 7.5 for an iron concentration of 0.01 M. One can see from the plot that k_1 is 97 L mol^{-1} s^{-1} and k_2 is 15.875/K_{a1} which is 7.8 × 10^3 L mol^{-1} s^{-1} using K_{a1} equal to 2.04 × 10^{-3}. The straight line lends support for the suggested mechanism in Box 7.1. In the same way a plot of k_r against (1/[H$^+$]) should give a straight line of slope $k_{-2}K_{a2}$ and intercept k_{-1}. The plot is shown in Fig. 7.6

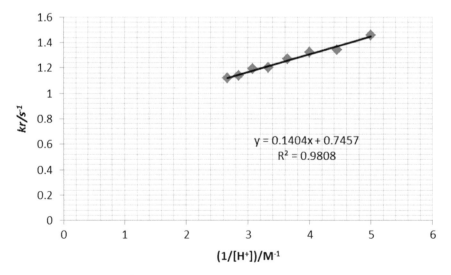

Fig. 7.6 The plot of k_r/s^{-1} against the reciprocal acid concentration for the iron(III) thiocyanate reaction at an iron concentration of 0.01 M after Mieling and Pardue [19]

Table 7.1 Stopped-flow kinetic data across three studies at $\mu = 1$ M for the mechanism in Box 7.1

Year	Author/s	k_1	k_{-1}	k_2	k_{-2}
1978	Mieling and Pardue	97 ± 3	0.75 ± 0.03	$(9.6 \pm 5) \times 10^3$	$(2.2 \pm 0.1) \times 10^3$
1997	Clark	109 ± 10	0.79 ± 0.1	$(8.2 \pm 0.8) \times 10^3$	$(2.63 \pm 0.23) \times 10^3$
2005	Hoag	89	0.72	6.9×10^3	–

from which one can see that k_{-1} is 0.75 s^{-1} and k_{-2} is $0.1404/K_{a2}$ which yields a k_{-2} value of 2.2×10^3 s^{-1} using a K_{a2} value of 6.5×10^{-5}. This means that Reaction 2 in Box 7.1 is more kinetically favoured but less thermodynamically favoured than Reaction 1. Kinetics is determined by the rate constants and thermodynamics by the equilibrium constants. The rate constant for the forward reaction divided by the rate constant for the reverse reaction will yield an equilibrium constant. The equilibrium constant for Reaction 1 calculates to be 129 M^{-1} and for Reaction 2, 3.5 M^{-1}.

A comparison of the kinetic stopped-flow data at an ionic strength of μ equal to 1 M for three studies using computable stopped-flow spectrophotometry for the mechanism in Box 7.1 is given in Table 7.1. A reasonable correspondence exists across the three data sets confirming the kinetic and thermodynamic differences between reactions (7.1) and (7.2) for the mechanism in Box 7.1.

7.3 Hydrogen Ion Independent One-Step Mechanism

In 2002, a stopped-flow kinetics experiment involving the iron(III) thiocyanate reaction was published online by Wayne Steinmetz at Pomona College [20]. This experiment considered two different mechanisms each different from that shown in Box 7.1. The experimental details suggest that both mechanisms lead mathematically to the same pseudo first-order rate law as that shown in Eq. (7.9), $\ln(A_{eq}/(A_{eq} - A)) = k_{obs}t$. However, the composition of k_{obs} is different in each case and the assumptions leading to Eq. (7.9) are important chemically and mathematically. The two mechanisms considered in the experiment are not hydrogen ion dependent as is the mechanism in Box 7.1. We examine the additional two mechanisms here to show how the same pseudo first-order rate law applies across all three mechanisms and how the composition of k_{obs} differs in each case.

Box 7.3 A one-step mechanism proposed for the formation of pentaaquathiocyanatoiron(III)

$$Fe(H_2O)_6^{3+} + SCN^- \underset{k_{-1}}{\overset{k_1}{\rightleftharpoons}} Fe(H_2O)_5SCN^{2+}$$

The first mechanism is the one-step mechanism shown in Box 7.3. One can write the corresponding rate law, removing charges and coordinated water for simplicity, as:

$$\frac{d[FeSCN]}{dt} = k_1[Fe][SCN] - k_{-1}[FeSCN] \tag{7.10}$$

[SCN] can be written as:

$$[SCN] = [SCN]_0 - [FeSCN] \tag{7.11}$$

where $[SCN]_0$ is the thiocyanate concentration at zero time. Substituting (7.11) into (7.10) gives:

$$\begin{aligned}
\frac{d[FeSCN]}{dt} &= k_1[Fe]\{[SCN]_0 - [FeSCN]\} - k_{-1}[FeSCN] \\
&= k_1[Fe][SCN]_0 - k_1[Fe][FeSCN] - k_{-1}[FeSCN] \\
&= k_1[Fe][SCN]_0 - [FeSCN]\{k_1[Fe] + k_{-1}\}
\end{aligned} \tag{7.12}$$

Since $d[FeSCN]_{eq}/dt = 0$ at equilibrium, then,

$$k_1[Fe]_{eq}[SCN]_0 = [FeSCN]_{eq}\{k_1[Fe]_{eq} + k_{-1}\} \tag{7.13}$$

Under circumstances where $[Fe]_0 \gg [SCN]_0$, then $[Fe]_0 \approx [Fe]_{eq} \approx [Fe] \approx [Fe]_T$ where $[Fe]_T$ is the total iron concentration. Equation (7.12) can now be written, in the light of Eq. (7.13) as:

$$\begin{aligned}
\frac{d[FeSCN]}{dt} &= [FeSCN]_{eq}\{k_1[Fe]_T + k_{-1}\} - [FeSCN]\{k_1[Fe]_T + k_{-1}\} \\
&= (k_1[Fe]_T + k_{-1})([FeSCN]_{eq} - [FeSCN])
\end{aligned}$$

Rearranging for integration gives:

$$\int_{[FeSCN]_0}^{[FeSCN]_t} \left\{ \frac{1}{[FeSCN]_{eq} - [FeSCN]} \right\} d[FeSCN] = \int_0^t k_{obs}dt \text{ where } k_{obs} = (k_1[Fe]_T + k_{-1}).$$

On integrating one gets:

$\ln \frac{[\text{FeSCN}]_{eq}}{[\text{FeSCN}]_{eq} - [\text{FeSCN}]} = k_{obs}t$: and, in terms of absorbances:

$\ln \frac{A_{eq}}{A_{eq} - A} = k_{obs}t$ which is the same as Eq. (7.9).

7.4 Hydrogen Ion Independent Two-Step Mechanism

The second mechanism is the two-step mechanism shown in Box 7.4. From Box 7.4 the rate of production of the complex, removing charges for simplicity, can be written as:

$$\frac{d[\text{Fe}(\text{H}_2\text{O})_5\text{SCN}]}{dt} = k_2[\text{Fe}(\text{H}_2\text{O})_5][\text{SCN}] - k_{-2}[\text{Fe}(\text{H}_2\text{O})_5\text{SCN}] \qquad (7.14)$$

Box 7.4 Mechanism showing a dissociation first step and an association second step

$\text{Fe}(\text{H}_2\text{O})_6^{3+} \underset{k_{-1}}{\overset{k_1}{\rightleftharpoons}} \text{Fe}(\text{H}_2\text{O})_5^{3+} + \text{H}_2\text{O} : k_1 = $ slow step, $k_{-1} = $ fast step

$\text{Fe}(\text{H}_2\text{O})_5^{3+} + \text{SCN}^- \underset{k_{-2}}{\overset{k_2}{\rightleftharpoons}} \text{Fe}(\text{H}_2\text{O})_5\text{SCN}^{2+} : k_2 = $ fast step, $k_{-2} = $ slow step

---

One can see from Box 7.4 that $\text{Fe}(\text{H}_2\text{O})_5^{3+}$ is an intermediate and if one assumes that this species concentration remains small and constant, then one can apply the steady-state approximation as follows:

$$\frac{d[\text{Fe}(\text{H}_2\text{O})_5]}{dt} = k_1[\text{Fe}(\text{H}_2\text{O})_6] - k_{-1}[\text{Fe}(\text{H}_2\text{O})_5]$$

$$- k_2[\text{Fe}(\text{H}_2\text{O})_5][\text{SCN}] + k_{-2}[\text{Fe}(\text{H}_2\text{O})_5\text{SCN}] = 0$$

Rearranging: $k_1[\text{Fe}(\text{H}_2\text{O})_6] - [\text{Fe}(\text{H}_2\text{O})_5]\{k_{-1} + k_2[\text{SCN}]\}$

$$+ k_{-2}[\text{Fe}(\text{H}_2\text{O})_5\text{SCN}] = 0$$

Therefore: $\left[\text{Fe}(\text{H}_2\text{O})_5\right] = \dfrac{k_1[\text{Fe}(\text{H}_2\text{O})_6] + k_{-2}[\text{Fe}(\text{H}_2\text{O})_5\text{SCN}]}{k_{-1} + k_2[\text{SCN}]} \qquad (7.15)$

Given the relative size of the rate constants proposed in the mechanism of Box 7.4, Eq. (7.15) can be simplified to: $[\text{Fe}(\text{H}_2\text{O})_5] \approx k_1\left[\text{Fe}(\text{H}_2\text{O})_6\right]/k_{-1}$

Equation (7.14) can now be written as:

$$\frac{d[Fe(H_2O)_5SCN]}{dt} = \frac{k_1k_2}{k_{-1}}[Fe(H_2O)_6][SCN] - k_{-2}[Fe(H_2O)_5SCN]$$

Substituting Eq. (7.11) for [SCN] and substituting $[Fe]_T$ for iron species as previously gives:

$$\frac{d[Fe(H_2O)_5SCN]}{dt} = \frac{k_1k_2}{k_{-1}}[Fe]_T\{[SCN]_0 - [Fe(H_2O)_5SCN]\} - k_{-2}[Fe(H_2O)_5SCN]$$

$$= \frac{k_1k_2}{k_{-1}}[Fe]_T[SCN]_0 - [Fe(H_2O)_5SCN]\left\{\frac{k_1k_2}{k_{-1}}[Fe]_T + k_{-2}\right\}$$

$$(7.16)$$

At equilibrium, $\frac{d[Fe(H_2O)_5SCN]_{eq}}{dt} = 0$, so,

$$\frac{k_1k_2}{k_{-1}}[Fe]_T[SCN]_0 = [Fe(H_2O)_5SCN]_{eq}\left\{\frac{k_1k_2}{k_{-1}}[Fe]_T + k_{-2}\right\}.$$

Equation (7.16) now becomes:

$$\frac{d[Fe(H_2O)_5SCN]}{dt} = \{\frac{k_1k_2}{k_{-1}}[Fe]_T + k_{-2}\}\{[Fe(H_2O)_5SCN]_{eq} - [Fe(H_2O)_5SCN]\}$$

Integrating:

$$\int_{[Fe(H_2O)_5SCN]_0}^{[Fe(H_2O)_5SCN]_t} \left\{\frac{1}{[Fe(H_2O)_5SCN]_{eq} - [Fe(H_2O)_5SCN]}\right\} d[Fe(H_2O)_5SCN]$$

$$= \int_0^t k_{obs}dt,$$

where

$$k_{obs} = \left\{\frac{k_1k_2}{k_{-1}}[Fe]_T + k_{-2}\right\}.$$

This leads to: $\ln\frac{[Fe(H_2O)_5SCN]_{eq}}{[Fe(H_2O)_5SCN]_{eq}-[Fe(H_2O)_5SCN]} = k_{obs}.t$, and $\ln\frac{A_{eq}}{A_{eq}-A} = k_{obs}.t$, which is the same as Eq. (7.9).

All three mechanisms shown in Boxes 7.1, 7.3 and 7.4 result in the same first-order rate law,

Table 7.2 A summary of the composition of k_{obs} for the mechanisms shown in Boxes 7.1, 7.3, and 7.4

Mechanism	k_{obs}	k_f	k_r	k_f/k_r
H$^+$ dependent, Box 7.1	$k_{-1} + \frac{k_{-2}K_{a2}}{[H]} + [Fe]_T \left(k_1 + \frac{k_2 K_{a1}}{[H]}\right)$	159	1.32	120.5
One-step Box 7.3	$(k_1[Fe]_T + k_{-1})$	154	1.4	110
Two-step Box 7.4	$\left\{\frac{k_1 k_2}{k_{-1}}[Fe]_T + k_{-2}\right\}$	154 k_{-1}	1.4 k_{-1}	110

$$\ln \frac{A_{eq}}{A_{eq} - A} = k_{obs}.t,$$

with the composition of k_{obs} shown in Table 7.2. If one considers the data from Mieling and Pardue [19] for 0.25 M acid, values for k_f, k_r and k_f/k_r can be calculated for the three mechanisms and these are also shown in Table 7.2. Given the fact that all three mechanisms lead to the same pseudo first-order rate law and similar k_f/k_r values, how can one decide which of the three mechanisms applies to the iron(III) thiocyanate reaction. Boxes 7.1 and 7.3 mechanisms basically assume the existence of a 7-coordinate activated complex and Box 7.4 mechanism involves a 5-coordinate intermediate and a 6-coordinate activated complex. Below et al. [12] considered these two possibilities and concluded that there was no evidence then available that could distinguish between the two possibilities. One would expect entropies of activation for a 7-coordinate activated complex to be more negative than for a 6-coordinate activated complex but I am not aware if this matter has been resolved.

In the case where the Fe^{3+} concentration exceeds the SCN$^-$ concentration the published literature favours the mechanism in Box 7.1. Whether this is the case when the concentration of SCN$^-$ exceeds the Fe^{3+} concentration has not been determined.

7.5 Simple Mechanism for the Production of Two Iron(III) Thiocyanate Complexes

As far as I know a mechanism for the reaction under high thiocyanate concentrations has not been published. Under these circumstances one would have to allow for the presence of Fe(H$_2$O)$_4$(SCN)$_2^+$ in addition to Fe(H$_2$O)$_5$SCN^{2+}, both species contributing to the blood-red colour. An application of the global analysis software, *ReactLab Kinetics* [21], to the absorbance data used to determine the equilibrium constants reported in Chap. 6 might serve as a preliminary example of how one might determine rate constants for the simple model shown in Box 7.5. Absorbance measurements are not taken at just one wavelength as mentioned in Chap. 6 but in the example given here, absorbance measurements were taken over 300 wavelengths between 400 and

700 nm and for forty times between 0 and 60 ms using the stopped-flow apparatus for an iron concentration of 1.5×10^{-4} M and a thiocyanate concentration of
0.25 M. That is, around 12,000 absorbance readings were used to determine the rate
constants. The rate constants are varied until the calculated absorbances match the
experimental absorbances as close as possible by minimizing the square of the difference between calculated and experimental absorbances. An example of the match
between the calculated and experimental absorbances is shown in Fig. 7.7 for wavelengths 475, 550, and 625 nm. The calculated rate constants are shown in Table 7.3 as
is the kinetically and thermodynamically determined equilibrium constants. One can
see that the simple mechanistic model has led to very good fitting of the absorbance
data and a very good correspondence between the kinetically and thermodynamically
determined equilibrium constants. Further research will need to be conducted on the
kinetics of formation of the second iron(III) thiocyanate complex, $Fe(H_2O)_4(SCN)_2^+$.
Are other models than the simple one suggested here more relevant to its chemistry?

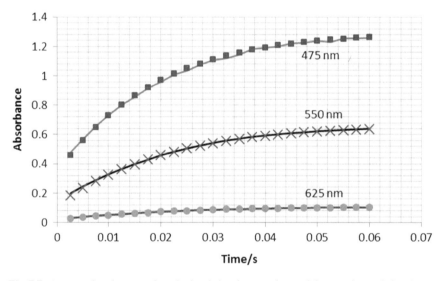

Fig. 7.7 A comparison between the calculated absorbance points and the experimental absorbance
lines for three wavelengths selected for the reaction between 1.5×10^{-4} M Fe^{3+} and 0.25 M SCN^-
at an ionic strength $\mu = 1.0$ M

Table 7.3 Rate constants obtained by global analysis of absorbance data like that shown in Fig. 7.7
at an ionic strength of 1 M for the simple mechanism in Box 7.5 and a comparison of equilibrium
constants calculated from kinetic and thermodynamic data

k_1/L^2 mol^{-1} s^{-1}	k_{-1}/s^{-1}	k_2/L^2 mol^{-1} s^{-1}	k_{-2}/s^{-1}	K_1 (kinetic)	K_1 (therm)	K_2 (kinetic)	K_2 (therm)
295	3.85	209.2	27.85	76.6	78.2	7.5	7.7

Is hydrogen ion dependence only applicable when $[Fe^{3+}]$ is in great excess or does it also apply when $[SCN^-]$ is in great excess?

Box 7.5 A simple mechanism showing the production of two complexes contributing to the blood-red colour for relatively high thiocyanate concentrations

$$Fe(H_2O)_6^{3+} + SCN^- \underset{k_{-1}}{\overset{k_1}{\rightleftharpoons}} Fe(H_2O)_5SCN^{2+} + H_2O$$

$$Fe(H_2O)_5SCN^{2+} + SCN^- \underset{k_{-2}}{\overset{k_2}{\rightleftharpoons}} Fe(H_2O)_4(SCN)_2^+ + H_2O$$

I have included some detail in the kinetic derivations here so one can appreciate how important assumptions are in the mathematics. This was also important in the derivations for the expressions for the equilibrium constants in an earlier chapter. By the beginning of the 19th century Newton had already made a major impact in physics due partly to his application of calculus to physical phenomena. While some were critical of introducing the theory of gravitational attraction between bodies in space and the concept of the infinitesimal in calculus, these ideas were so successful that chemists pondered whether such power was possible in a field like chemistry. As it turned out it was in the fields of thermodynamics and kinetics that chemistry came into its own towards the end of the 19th century and early in the 20th century. And it was chemists like Esson who had some proficiency in calculus who were able to combine with others like Harcourt who lacked this proficiency to produce valuable research in kinetics mentioned earlier. When different mechanisms lead mathematically to the same rate law as illustrated in this chapter one is forced to consider other properties which might lend a clue as to which mechanism is preferred but this is not always an easy task.

References

1. Laidler KJ (1993) The world of physical chemistry. Oxford University Press, Oxford
2. Wilhelmy L (1850) Über das Gesetz, nach welchem die Einwirkung der Säuren auf den Rohrzucker stattfinden. Poggendorff's Ann der Phys Und Chem 81(2):413–433, 499–526
3. Bertholet M, de Saint-Gilles PL (1862) Recherches sur les affinites: De la formation et de la décomposition des éthers. Ann De chim et de pharm 65:385–422
4. Harcourt AV, Esson W (1865) On the laws of connexion between the conditions of a chemical change and its amount. Phil Trans 156:193–222
5. Harcourt AV, Esson W (1895) On the laws of connexion between the conditions of a chemical change and its amount. Part III. Further researches on the reaction of hydrogen dioxide and hydrogen iodide. Phil Trans A 186:817–895
6. Van't Hoff JH (1884) Études de dynamique chimique. F. Muller, Amsterdam
7. Laidler KJ (1993) The world of physical chemistry. Oxford University Press, Oxford, p 236

8. Waage P, Guldberg CM (1864) Forhanlinger: Videnskabs-Selskabet I Christiana (trans: Studies concerning affinity by Abrash HI). In: Bastiansen O (ed) The law of mass action, a centenary volume. Universitetsforlaget, Oslo
9. Quílez J (2018) A historical/epistemological account of the foundation of the key ideas supporting chemical equilibrium theory. Found Chem. https://doi.org/10.1007/s10698-018-9320-0
10. Laidler KJ (1993) The world of physical chemistry. Oxford University Press, Oxford, p 234
11. Quílez J (2018) A historical/epistemological account of the foundation of the key ideas supporting chemical equilibrium theory. Found Chem 23. https://doi.org/10.1007/s10698-018-9320-0
12. Below JF, Connick RE, Coppel CP (1958) Kinetics of the formation of the ferric thiocyanate complex. J Am Chem Soc 80:2961–2967
13. Goodall DM, Harrison PW, Hardy MJ, Kirk CJ (1972) Relaxation kinetics of ferric thiocyanate. J Chem Educ 49(10):675–678
14. Gray ET, Workman HJ (1980) An easily constructed and inexpensive stopped-flow system for observing rapid reactions. J Chem Educ 57(10):752–755
15. Clark CR (1997) A stopped-flow kinetics experiment for advanced undergraduate laboratories: formation of iron(III) thiocyanate. J Chem Educ 74(10):1214–1217
16. de Berg K, Maeder M, Clifford S (2016) A new approach to the equilibrium study of iron(III) thiocyanates which accounts for the kinetic instability of the complexes particularly observable under high thiocyanate concentrations. Inorg Chim Acta 445:155–159
17. de Berg K, Maeder M, Clifford S (2017) The thermodynamic formation constants for iron(III) thiocyanate complexes at zero ionic strength. Inorg Chim Acta 446:249–253
18. Hoag CM (2005) Simple and inexpensive computer interface to a durrum stopped-flow apparatus tested using the iron(III)-thiocyanate reaction. J Chem Educ 82(12):1823–1825
19. Mieling GE, Pardue HL (1978) Evaluation of a computer-controlled stopped-flow system for fundamental kinetic studies. Anal Chem 50(9):1333–1337
20. https://pages.pomona.edu/~wsteinmetz/chem160/stopflow_2002.doc
21. http://www.jplusconsulting.com/

Chapter 8
The Reaction in Secondary and Tertiary Education Curricula

8.1 Course Work and Laboratory Work

What makes the iron(III) thiocyanate reaction of benefit to chemistry educators is the fact that its K value for an ionic strength of 0.5 M is neither too high nor too small which means its incomplete nature should be more noticeable when discussing chemical equilibrium. Quílez [1] also suggests that the iron(III) thiocyanate reaction is ideal for discussing the incompleteness of a reaction at equilibrium. The iron(III) thiocyanate reaction arguably entered the educational folklore of chemical equilibrium with the introduction of the CHEM STUDY curriculum for secondary school students in 1960 in the USA. The reaction in the form: $Fe^{3+}(aq) + SCN^-(aq) \rightleftharpoons FeSCN^{2+}(aq)$, was used in the textbook [2] to discuss the impact of concentration changes on the position of equilibrium by observing changes in the intensity of the red coloured $FeSCN^{2+}$ species. The laboratory manual [3] challenged students to determine the expression for the equilibrium constant for the reaction by a visual comparison of colour intensities. It is true that the reaction featured in articles earlier than 1960 in the *Journal of Chemical Education* (for example, Lewin and Wagner [4]) and in Fowles' *Lecture Experiments in Chemistry* [5], but the frequency with which it has been studied and published in the *Journal of Chemical Education* increased from one before 1960 to thirteen subsequent to 1960 according to Table 1.1. It is interesting to note that the reverse is true for the reaction's presence in chemistry research journals, with an estimated twenty papers before 1960 and ten papers subsequent to 1960 according to Table 1.1. Secondary chemistry textbooks before CHEM STUDY tended to use the reaction in chemical analysis rather than for equilibrium study. Barrell [6] used a qualitative analysis section to distinguish between Fe^{3+}, Al^{3+}, and Cr^{3+}. The test for iron is given as the production of the deep red colour of $Fe(CNS)_3$ on adding nitric acid and potassium thiocyanate. The other ions do not undergo this reaction.

The current secondary Higher School Certificate (HSC) Chemistry Syllabus in New South Wales, Australia, does not specify the use of the iron(III) thiocyanate reaction when dealing with chemical equilibrium in the theory section or in the

© The Author(s), under exclusive license to Springer Nature Switzerland AG 2019
K. C. de Berg, *The Iron(III) Thiocyanate Reaction*, SpringerBriefs in
History of Chemistry, https://doi.org/10.1007/978-3-030-27316-3_8

recommended laboratory activities [7]. The focus is on the Arrhenius and Brónsted-Lowry acid-base equilibria in a qualitative sense and students are only introduced to equilibrium calculations if they choose the Industrial Chemistry elective. This is in stark contrast to the CHEM STUDY course where equilibrium calculations were part of the core activity of the course. A number of reasons may be advanced for this difference in emphasis: HSC chemistry caters for a broader spectrum of students than did CHEM STUDY and this is reflected in a reduction of mathematical content in HSC; HSC chemistry purposely sets out to include content from the history and nature of chemistry, which led to a reduction in some of the traditional content such as an emphasis on quantitative problem solving; and more is known about the history of acid-base chemistry than the iron(III) thiocyanate reaction. However, the formation of the blood-red colour from the addition of a colourless acidified iron(III) salt to a colourless thiocyanate salt in aqueous solution makes the iron(III) thiocyanate reaction very suitable for chemical analysis and equilibrium study, if one acknowledges the caution previously suggested when explaining the changes in colour on adding more thiocyanate (Chap. 4). One of the purposes of this study has been to provide an informative account of the reaction's history to fill the gap in our current knowledge of this history. Another purpose of this study has been to show that an historical approach to chemistry need not come at the expense of its mathematical content.

A lack of historical content in chemistry education can lead to what Schwab [8, 9] called a *rhetoric of conclusions*. Concepts relevant to chemical reaction behaviour such as those elaborated by Gladstone [10] for complete or incomplete transformation develop and change over time and this development gives us access to the nature of chemistry, an important ingredient of scientific literacy. That historical study has a bearing on current chemical literacy is borne out by the *Rationale* statement of the New South Wales HSC Stage 6 Chemistry Syllabus [11] as follows:

> The History and Philosophy of Science (HPS) as it relates to the development of the understanding, utilisation and manipulation of chemical systems is important in developing current understanding in Chemistry and its related applications in the context of technology, society and the environment.

That HPS and nature of science (NOS) are closely related can be seen in how the issues and topics of the journal, *Science & Education-Contributions from History, Philosophy and Sociology of Science and Mathematics*, developed from the journal's inception in 1992. Within five years of the journal's first publication two thematic issues on the nature of science were published with NOS being addressed in some form almost every year since that time.

A technique I have found useful when introducing students to science history is the use of role play [12]. In this reference, I relate how I revised some copper chemistry with my students (Year 12 or first-year university) by setting up a dialogue between Joseph Priestley, played by the teacher or lecturer, and an Italian chemist, Romeo Pedanti, played by a student and who is a recent convert to Lavoisier's chemistry. A piece of copper sheet is heated lightly in a flame and then strongly in the flame. A debate ensues about what causes the changes to the copper on heating.

The point of the role play is to introduce students to the Priestley/Lavoisier debate about combustion. Now a similar role play and debate could be established between Bergman and Berthollet about whether the iron(III) thiocyanate reaction goes to completion or not. Good arguments could be presented from both sides given the detail provided in Chap. 4.

The Salters Advanced Chemistry Course in England [13] includes the iron(III) thiocyanate reaction as an equilibrium activity explaining changes to the equilibrium as Fe^{3+} and SCN^- concentrations are increased and decreased. The chromate-dichromate equilibrium is also included prior to the iron(III) thiocyanate equilibrium. Whether this was intended or not it makes sense to study these two equilibria in this order. Because chromate has a different colour to dichromate it is easier to justify a reverse reaction, when looking at the colour change on adding hydroxide ions for example. Because only the complex in the iron(III) thiocyanate reaction is coloured, one cannot really tell if the increase in colour intensity on adding more thiocyanate is due to residual Fe^{3+} reacting to form more complex or whether the $FeSCN^{2+}$ reacts with added SCN^- to form $Fe(SCN)_2^+$ which we know exists and has a deep red colour as well.

General Chemistry textbooks suitable for teaching chemistry to college and university students at the first-year level (for example, [14–16]) generally do not discuss the iron(III) thiocyanate reaction in the equilibrium chapter because the chapter is largely dedicated to gas phase equilibria. However, the reaction is often featured in the laboratory manuals since it is more easily studied than gas phase equilibria. The college chemistry laboratory manual by Hein et al. [17] uses the iron(III) thiocyanate equilibrium to apply Le Chatelier's principle to explain what happens when solid iron(III) chloride and solid potassium thiocyanate are added to separate samples of the equilibrium. The addition of silver nitrate to another sample of the equilibrium is designed to challenge students. The General Chemistry laboratory manual by Beran [18] asks students to use 0.2 M $Fe(NO_3)_3$ in 0.1 M HNO_3 and 0.001 M NaSCN in 0.1 M HNO_3 to construct a standard curve and to use this to determine the equilibrium compositions in five mixtures of iron(III) and thiocyanate and hence to calculate the equilibrium constants. Five 25 mL standard solutions were prepared containing 0.08 M Fe^{3+} and 0 M, 4×10^{-5} M, 8×10^{-5} M, 12×10^{-5} M and 16×10^{-5} M SCN^- in HNO_3. The significance of these reagent concentrations has been referred to previously. The absorbance of these five solutions was measured with a spectrophotometer and plotted against the $[FeSCN^{2+}]$ which is effectively the $[SCN^-]$ as the $[Fe^{3+}]$ is in large excess. This is an important assumption given the equilibrium constant has not yet been determined. A set of five 10 mL solutions all containing 5 mL of 0.002 M Fe^{3+} in 0.1 M HNO_3 and 1 mL, 2 mL, 3 mL, 4 mL, and 5 mL of 0.002 M SCN^- in 0.1 M HNO_3 and all diluted to the 10 mL mark with 0.1 M HNO_3 were prepared and the absorbances measured. Using the standard curve, the $[FeSCN^{2+}]$ was determined for each solution and the equilibrium constant calculated.

The Physical Chemistry laboratory manual by James and Prichard [19] uses 0.002 M $Fe(NO_3)_3$ and 0.002 M KSCN in acid solutions to apply Job's method of continuous variations to determine the formula of the iron(III) thiocyanate complex. This technique was explained in an earlier chapter.

8.2   The Use of Excel Spreadsheets at the Tertiary Level

The reaction also features in the study of chemometrics at the tertiary level. The purpose of the exercise explained in Chap. 3 of Maeder and Neuhold [20] is to show how an Excel spreadsheet can be used to calculate concentration profiles for Fe^{3+}, SCN^-, and $FeSCN^{2+}$ for a titration of 10 mL of 0.1 M Fe^{3+} with 9×10^{-2} M SCN^-. The equilibrium is written as follows.

$$Fe^{3+} + SCN^- \overset{K}{\rightleftharpoons} Fe(SCN)^{2+}$$

The associated equilibrium constant expression using what has historically been called the law of mass action is:

$$K = \frac{[Fe(SCN)^{2+}]}{[Fe^{3+}][SCN^-]} \tag{8.1}$$

One can express the total concentrations of iron and thiocyanate as follows:

$$[Fe^{3+}]_{total} = [Fe^{3+}] + \left[Fe(SCN)^{2+}\right]$$
$$[SCN^-]_{total} = [SCN^-] + \left[Fe(SCN)^{2+}\right] \tag{8.2}$$

Substituting the two equations in (8.2) into Eq. (8.1) yields:

$$K = \frac{\left[Fe(SCN)^{2+}\right]}{([Fe^{3+}]_{total} - \left[Fe(SCN)^{2+}\right])([SCN^-]_{total} - \left[Fe(SCN)^{2+}\right])} \tag{8.3}$$

If K is known, the use of the equations in (8.2) has reduced an expression containing three variables (Eq. 8.1) to an equation containing one variable (Eq. 8.3). Expanding out the brackets in Eq. (8.3) and rearranging produces a quadratic equation in $[Fe(SCN)^{2+}]$.

$$K\left[Fe(SCN)^{2+}\right]^2 - \left\{K\left([Fe^{3+}]_{total} + [SCN^-]_{total}\right) + 1\right\}$$
$$\left[Fe(SCN)^{2+}\right] + K[Fe^{3+}]_{total}[SCN^-]_{total} = 0 \tag{8.4}$$

Using the quadratic formula, this leads to the following solution:

$$[Fe(SCN)^{2+}]$$

$$= \frac{\{K([Fe^{3+}]_{total} + [SCN^-]_{total}) + 1\} - \sqrt{\{K([Fe^{3+}]_{total} + [SCN^-]_{total}) + 1\}^2 - 4K^2[Fe^{3+}]_{total}[SCN^-]_{total}}}{2K}$$

$$(8.5)$$

Having found the $[Fe(SCN)^{2+}]$, one can use the equations in (8.2) to find $[Fe^{3+}]$ and $[SCN^-]$ and process the calculations on an Excel spreadsheet. A sample of the spreadsheet is shown in Fig. 8.1. The Excel formulae used for the concentration columns is shown in Table 8.1. The concentration profiles for a K value of 200 are shown in Fig. 8.2. One can see that after 25 mL of SCN^- has been added, the blood-

K	200					
V_0	0.01					
[Fe3+]_0	0.1					
[SCN-]_add	0.09					
V_added	V_tot	[Fe3+]_tot	[SCN-]_tot	[Fe(SCN)2+]	[Fe3+]	[SCN-]
0	0.01	0.1	0	0	0.1	0
0.001	0.011	0.0909091	0.0081818	0.00771795	0.083191	0.000464
0.002	0.012	0.0833333	0.015	0.01399115	0.069342	0.001009
0.003	0.013	0.0769231	0.0207692	0.01911582	0.057807	0.001653
0.004	0.014	0.0714286	0.0257143	0.02329453	0.048134	0.00242
0.005	0.015	0.0666667	0.03	0.02666667	0.04	0.003333
0.006	0.016	0.0625	0.03375	0.02932909	0.033171	0.004421
0.007	0.017	0.0588235	0.0370588	0.0313524	0.027471	0.005706
0.008	0.018	0.0555556	0.04	0.03279542	0.02276	0.007205
0.009	0.019	0.0526316	0.0426316	0.03371793	0.018914	0.008914
0.01	0.02	0.05	0.045	0.03418861	0.015811	0.010811

Fig. 8.1 Excel Spreadsheet data for a titration where 0.09 M SCN^- is added, 1 mL at a time, to 10 mL of 0.1 M Fe^{3+}. The concentration of the species in the iron(III) thiocyanate equilibrium is shown in the last three columns. The Excel columns go from A to G and the rows go from 1 to 17

Table 8.1 Excel formulae used for finding the concentrations pertinent to the iron(III) thiocyanate equilibrium

Species	Excel formula
$[Fe^{3+}]_{tot}$	=$B$3 * $B$2/B7
$[SCN^-]_{tot}$	=$B$4 * A7/B7
$[Fe(SCN)^{2+}]$	=(($B$1 * (C7 + D7) + 1) − SQRT(($B$1 * (C7 + D7) + 1)^2 − 4 * $B$1^2 * C7 * D7))/(2 * $B$1)
$[Fe^{3+}]$	=(C7 − E7)
$[SCN^-]$	=(D7 − E7)

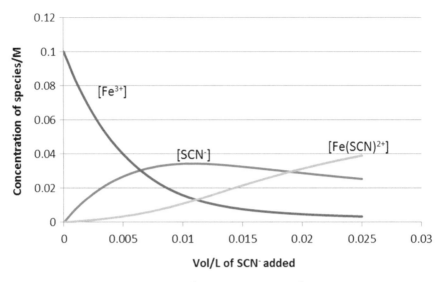

Fig. 8.2 The concentration profiles for Fe^{3+}, SCN^-, and $Fe(SCN)^{2+}$ as 0.09 M SCN^- is added to 10 mL of 0.1 M Fe^{3+}. An equilibrium constant, K, of 200 was used in the calculations

red complex, $Fe(SCN)^{2+}$, has the highest concentration of the three species. This is an exercise designed to show students the power of Excel as a data processing tool when combined with some mathematics. The problem mentioned in Chap. 4 about the instability of the blood-red complex has not been accounted for here as the overall purpose of the exercise was experience in data handling rather than determining an accurate value for the equilibrium constant.

A more advanced application of this spreadsheet method is to measure the absorbance of each mixture and compare with the calculated absorbance which can be found by multiplying the formula for $[Fe(SCN)^{2+}]$ by the molar absorptivity, ε, for a 1 cm absorbance cell. One then chooses the *Solver* add-on in the Excel spreadsheet and uses it to allow K and ε to change so as to minimize the square of the differences between the experimental and calculated absorbances. The fitting works best if the chosen values of K and ε are not too different from the accepted values. I have mentioned this technique in Chap. 6.

A big advantage of using formulae to construct concentration profiles on a spreadsheet is that one can change the value of constants like K and look at the impact on the concentration profiles. Changing K to a very large number like 2×10^5 gives a concentration profile like that shown in Fig. 8.3. One can see that SCN^- is effectively acting as a limiting reagent up until just after the addition of about 11 mL of SCN^- so it is difficult to discuss the concept of *incomplete reaction* under these circumstances. In fact, the Fe^{3+} and SCN^- concentrations are zero at the 11 mL mark. Talking about incomplete reactions is hard to justify under these circumstances. At no point in Fig. 8.3 are all three species obviously present. If K is changed to a very small number like 2×10^{-5}, one gets the profiles in Fig. 8.4. At no point in the

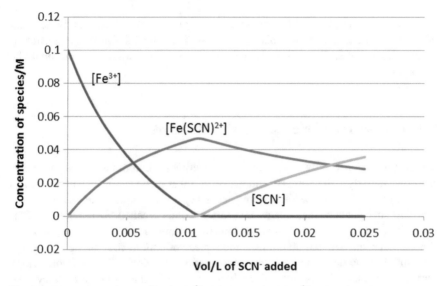

Fig. 8.3 The concentration profiles for Fe^{3+}, SCN^-, and $Fe(SCN)^{2+}$ as 0.09 M SCN^- is added to 10 mL of 0.1 M Fe^{3+}. An equilibrium constant, K, of 2×10^5 was used in the calculations

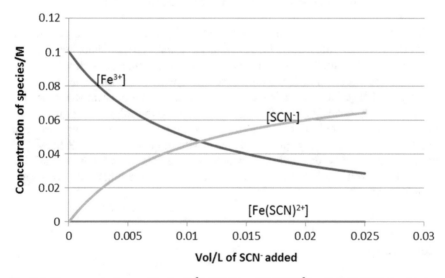

Fig. 8.4 The concentration profiles for Fe^{3+}, SCN^-, and $Fe(SCN)^{2+}$ as 0.09 M SCN^- is added to 10 mL of 0.1 M Fe^{3+}. An equilibrium constant, K, of 2×10^{-5} was used in the calculations

25 mL addition of thiocyanate are all three species present. In fact, K is so small that no measureable amounts of $Fe(SCN)^{2+}$ have formed. This is why the iron(III) thiocyanate equilibrium has attracted the attention of educators since its K is neither too large or too small so that all three species are present in the titration. The problem

educators need to be aware of is the issue of interpreting the outcome of the addition of SCN^- to the equilibrium. This was discussed in Chap. 4. Under these circumstances the use of an Excel spreadsheet to calculate concentration profiles could be important.

8.3 Commercial Science Activity Packages

Commercial science activity packages are sometimes incorporated into science curricula particularly at the primary school level or lower secondary level where teachers commonly lack the science background necessary for developing one's own curriculum. One such package for students ten years and older is called 'Artificial Blood Set' [21]. This package produces so-called blood by combining ferric chloride with potassium thiocyanate aqueous solutions using a piece of cotton wool on the arm. However, there are some difficulties in the explanations provided. The equation written for the explanation shows ammonium thiocyanate instead of potassium thiocyanate as follows:

$$FeCl_3 + NH_4SCN \rightarrow FeSCN^{2+} + NH_4Cl + 2Cl^{1-}$$

The blood-red colour is removed with sodium thiosulphate and the explanation given is, "Iron(III) Thiocyanate (aq) is oxidized. It reacts with a reducing solution-dilute Sodium Thiosulphate to become neutralized and form a colourless solution." If the sodium thiosulphate is a reducing solution, which it is, then one should say that the iron(III) thiocyanate is *reduced*. From a child's future chemistry education perspective, an interesting question emerges: "Do the thiosulphate ions react with residual ferric ions, reduce them to ferrous ions and thus force the reaction to occur in reverse, or, do the thiosulphate ions react with the complex ions, $FeSCN^{2+}$, to produce the complex ions, $FeSCN^+$, which must be colourless?" One explanation is consistent with Berthollet's *incomplete reaction*, and the other with Bergman's *complete reaction*.

References

1. Quílez J (2004) A historical approach to the development of chemical equilibrium through the evolution of the affinity concept: some educational suggestions. Chem Ed Res Prac 5(1):69–87
2. Pimentel GC (ed) (1963) CHEMISTRY: an experimental science. W.H. Freeman & Co., San Francisco
3. Malm LE(ed) (1963) Laboratory manual for chemistry (CHEM STUDY). W.H. Freeman & Co., San Francisco
4. Lewin SZ, Wagner RS (1953) The nature of iron(III) thiocyanate in solution. J Chem Educ 445–449
5. Fowles G (1937) Lecture experiments in chemistry. G. Bell & Sons Ltd., London

6. Barrell FT (1960) Complete chemistry for seniors. The Jacaranda Press, Brisbane
7. Board of Studies (2002) Chemistry Stage 6 syllabus. NSW Board of Studies, Sydney
8. Schwab JJ (1962) The teaching of science as enquiry. Harvard University Press, Cambridge
9. Schwab JJ (1974) The concept of the structure of a discipline. In: Eiser EW, Vallance E (eds) Conflicting conceptions of curriculum. McCutchan Pub. Corp., Berkeley, pp 162–175
10. Gladstone JH (1855) On circumstances modifying the action of chemical affinity. Phil Trans R Soc Lond 145:179–223
11. Board of Studies (2002) Chemistry Stage 6 syllabus. NSW Board of Studies, Sydney, p 6
12. Kilgour P, Hinz J, Petrie K, Long W, de Berg K (2015) Role-playing: a smorgasbord of learning types. Int J Innov Interdisc Res 3(1):11–24
13. University of York Science Education Group (1994) Salters advanced chemistry. Heineman, Oxford
14. Blackman A, Bottle S, Schmid S, Mocerino M, Wille U (2016) Chemistry, 3rd edn. Wiley, Milton
15. Silberberg M (2009) Chemistry: the molecular nature of matter and change, 5th edn. McGraw-Hill, New York
16. Burrows A, Holman J, Parsons A, Pilling G, Price G (2009) Chemistry[3]: introducing inorganic, organic and physical chemistry. Oxford University Press, Oxford
17. Hein M, Peisen JN, Miner RL (2011) Foundations of college chemistry in the laboratory, 13th edn. Wiley, Hoboken
18. Beran JA (2004) Laboratory manual for principles of general chemistry. Wiley, Hoboken
19. James AM, Prichard FE (1974) Practical physical chemistry, 3rd edn. Longman, London
20. Maeder M, Neuhold Y-M (2007) Practical data analysis in chemistry. Elsevier, Amsterdam
21. Science Time Series-Chemical Science (2007) Artificial blood set. Eastcolight Ltd., Hong Kong

Chapter 9
Conclusion

The practice of modern chemistry involves a particular way of viewing the world and the universe quite unlike that of a child or an adult not educated in modern chemistry. A chemist thinks of matter in terms of its abstraction into pure elements or compounds in order to apply the models, theories, and laws that have developed in the history of chemistry to help us understand the changes that occur in our world and in our bodies. One way of illustrating this is to show how our view of a chemical reaction changed over time as our tools became more sophisticated. Building a model of a chemical reaction proves somewhat more challenging than building a house, because the building materials for a chemical reaction are invisible and require special instruments to access them and understand their behaviour. The reaction between ferric ions and thiocyanate ions was first studied seriously in 1855 by Gladstone at a time of significant controversy about the nature of a chemical reaction. Forty-seven papers dealing with the iron(III) thiocyanate system from 1826 to 2017 have been studied and listed in Table 1.1 to help present an historical image of how this reaction has been understood and used over a span of 191 years. The reaction was seen by Waage and Guldberg as supporting their principle of chemical equilibrium largely through the interpretation of observed colour changes indicating *incomplete transformation*. However, chemistry educators have demonstrated that students have difficulty coming to this conclusion given their early chemistry background, where limiting reagents had to be identified in a chemical reaction so amounts of products could be accurately calculated from the balanced chemical equation assuming complete transformation of the limiting reagent. The idea of chemical equilibrium allowed for all reagents to be incompletely transformed. The narrative of how chemists made the transition from *complete* reactions to *incomplete* reactions; from *grain measures* to *grams*; from chemical *affinity* to *free energy*; and from $Fe_2O_3, 3NO_5$ to $Fe(NO_3)_3$ are some important highlights that can inform the nature of chemistry. And the context of the narrative reaches back into the 18th century and before to the concept of affinity and the nature of an exact science as shown in Chap. 1.

© The Author(s), under exclusive license to Springer Nature Switzerland AG 2019 97
K. C. de Berg, *The Iron(III) Thiocyanate Reaction*, SpringerBriefs in
History of Chemistry, https://doi.org/10.1007/978-3-030-27316-3_9

But what is clear is that our understanding of the iron(III) thiocyanate reaction is still undergoing change, as shown in Table 1.1 and discussed throughout this book, inspite of the great progress that has been made through the application of mathematics and data analysis. This is associated with the unresolved reason for the instability of the iron(III) thiocyanate complexes and the nature of the complexes at very high thiocyanate concentrations relative to iron(III) concentrations and very high iron(III) concentrations relative to thiocyanate concentrations. However, there is a growing consensus that for aqueous thiocyanate concentrations up to 0.25 M the species responsible for the blood-red colour are $Fe(H_2O)_5SCN^{2+}$ and $Fe(H_2O)_4(SCN)_2^+$ and the reason for the instability is most likely due to reduction of the Fe^{3+} ions to Fe^{2+} ions by thiocyanate ions. The most recent values for the equilibrium constants associated with the formation of these complexes [1, 2] (Table 1.1) are probably the most trustworthy given the use of initial spectra taken before the effects of instability become noticeable.

It has been shown that the laws of chemistry are often mathematical in nature and the iron(III) thiocyanate equilibrium is no exception. While mathematical equations are of value in quantitative analysis, the real value lies in the chemical assumptions used in deriving the final result. This appears to be of particular importance in the case of iron(III) thiocyanates. The coming of mathematics to chemistry is a fulfilment of the dream expressed by 18th century chemists who had to tread a difficult path in the shadow of victorious physics and astronomy. The calculus entered chemistry through classical thermodynamics and kinetics towards the end of the 19th century and these tools were used in uncovering the chemistry of the iron(III) thiocyanate reaction as summarised in Table 1.1. The so-called exactness and fundamental nature of physics was so strongly felt for a time that it was considered that chemistry and biology would ultimately be reduced to physics. Hettema [3] counters this thought by arguing that there are "sound philosophical reasons why we have to drop the normative picture of reduction (chemistry reduced to physics) and adopt a notion that is more amenable to how real theories of chemistry and physics work." In the case of transition state theory in kinetics, Hettema [4] shows how a network of theories drawn from both physics and chemistry contribute to an understanding of the nature of the transition state. It is in this sense that Hettema speaks about the union of physics and chemistry rather than the reduction of chemistry to physics.

A common concern in history and philosophy of science is whether experimental data leads uncritically to scientific laws and theories. In other words, does the absorbance data for the iron(III) thiocyanate system lead uncritically to the chemical model and its equilibrium data. In the section on mathematics and data analysis one can see that the chemical model was first assumed and given the model the outcome was predicted through the application of mathematics and/or data analysis on the chemical model. Even the equilibrium constants and molar absorbance data had to be predicted when global analysis was applied. In this sense one assumes the answer rather than deriving it from experimental data. While this appears to be a kind of circular reasoning it has an inbuilt filter in that the assumed model can be adjusted until the predicted outcome matches as closely as possible the experimental data. A question of philosophical significance is whether the accepted chemical model is unique. Is it possible that two different chemical models could lead to the same outcomes? This appears to be particularly the case in the study of kinetics where two different mechanisms can lead to the same rate law. This question is generally not addressed in the papers shown in Table 1.1. Neither is the idea of *law* as *description*, *theory* as *explanation*, and *model* as *representation* an emphasis in the papers of Table 1.1. Chemists generally tend not to meddle in history and philosophy of science in their chemistry publications [5], some exceptions being Berthelot and Mendeleev in the 19th century [6, 7], Partington in the 20th century [8], and Hoffman in the 21st century [9]. Is it time that this general trend in chemistry publications was allowed to change? The signs are hopeful given the interest chemists have shown in this series on the history of chemistry and molecular science.

References

1. de Berg K, Maeder M, Clifford S (2016) A new approach to the equilibrium study of iron(III) thiocyanates which accounts for the kinetic instability of the complexes particularly observable under high thiocyanate concentrations. Inorg Chim Acta 445:155–159
2. de Berg K, Maeder M, Clifford S (2017) The thermodynamic formation constants for iron(III) thiocyanate complexes at zero ionic strength. Inorg Chim Acta 446:249–253
3. Hettema H (2017) The union of chemistry and physics: linkages, reduction, theory nets and ontology. Springer, Switzerland, p 4
4. Hettema H (2017) The union of chemistry and physics: linkages, reduction, theory nets and ontology. Springer, Switzerland, p 82
5. Good RJ (1999) Why are chemists turned off by philosophy? Found Chem 1:65–96
6. Mendeleev DI (1889) The periodic law of the chemical elements. J Chem Soc 55:634–656
7. Bolton HC (1896) Berthelot's contributions to the history of chemistry. J Am Chem Soc 18(5):466–474
8. Partington JR (1937) A short history of chemistry. Macmillan, London
9. Hoffmann R (2012) Nearly circular reasoning. In: Kovac J, Weisberg M (eds) Roald Hoffmann on the philosophy, art, and science of chemistry. Oxford University Press, New York, pp 45–52

Printed in the United States
By Bookmasters